Deformation, Fracture and Microstructure of Metallic Materials

Deformation, Fracture and Microstructure of Metallic Materials

Editors

Xiao-Wu Li
Peng Chen

MDPI • Basel • Beijing • Wuhan • Barcelona • Belgrade • Manchester • Tokyo • Cluj • Tianjin

Editors
Xiao-Wu Li
School of Materials Science
and Engineering
Northeastern University
Shenyang
China

Peng Chen
School of Materials Science
and Engineering
Northeastern University
Shenyang
China

Editorial Office
MDPI
St. Alban-Anlage 66
4052 Basel, Switzerland

This is a reprint of articles from the Special Issue published online in the open access journal *Metals* (ISSN 2075-4701) (available at: www.mdpi.com/journal/metals/special_issues/deformation_fracture_microstructure).

For citation purposes, cite each article independently as indicated on the article page online and as indicated below:

LastName, A.A.; LastName, B.B.; LastName, C.C. Article Title. *Journal Name* **Year**, *Volume Number*, Page Range.

ISBN 978-3-0365-7863-7 (Hbk)
ISBN 978-3-0365-7862-0 (PDF)

© 2023 by the authors. Articles in this book are Open Access and distributed under the Creative Commons Attribution (CC BY) license, which allows users to download, copy and build upon published articles, as long as the author and publisher are properly credited, which ensures maximum dissemination and a wider impact of our publications.

The book as a whole is distributed by MDPI under the terms and conditions of the Creative Commons license CC BY-NC-ND.

Contents

Preface to "Deformation, Fracture and Microstructure of Metallic Materials" vii

Xiaowu Li and Peng Chen
Deformation, Fracture and Microstructure of Metallic Materials
Reprinted from: *Metals* 2023, 13, 1015, doi:10.3390/met13061015 1

Marcos Sánchez, Sergio Cicero, Mark Kirk, Eberhard Altstadt, William Server and Masato Yamamoto
Using Mini-CT Specimens for the Fracture Characterization of Ferritic Steels within the Ductile to Brittle Transition Range: A Review
Reprinted from: *Metals* 2023, 13, 176, doi:10.3390/met13010176 5

Xiaowu Li, Xianjun Guan, Zipeng Jia, Peng Chen, Chengxue Fan and Feng Shi
Twin-Related Grain Boundary Engineering and Its Influence on Mechanical Properties of Face-Centered Cubic Metals: A Review
Reprinted from: *Metals* 2023, 13, 155, doi:10.3390/met13010155 25

Hai Qiu and Tadanobu Inoue
Evolution of Poisson's Ratio in the Tension Process of Low-Carbon Hot-Rolled Steel with Discontinuous Yielding
Reprinted from: *Metals* 2023, 13, 562, doi:10.3390/met13030562 45

Jun Cao, Haidong Jia, Weifeng Ma, Ke Wang, Tian Yao and Junjie Ren et al.
Repair Reliability Analysis of a Special-Shaped Epoxy Steel Sleeve for Low-Strength Tee Pipes
Reprinted from: *Metals* 2022, 12, 2149, doi:10.3390/met12122149 59

Jiahui Du, Peng Chen, Xianjun Guan, Jiawei Cai, Qian Peng and Chuang Lin et al.
The Effect of Strain Rate on the Deformation Behavior of Fe-30Mn-8Al-1.0C Austenitic Low-Density Steel
Reprinted from: *Metals* 2022, 12, 1374, doi:10.3390/met12081374 73

Zhenlei Li, Rui Zhang, Dong Chen, Qian Xie, Jian Kang and Guo Yuan et al.
Quenching Stress of Hot-Rolled Seamless Steel Tubes under Different Cooling Intensities Based on Simulation
Reprinted from: *Metals* 2022, 12, 1363, doi:10.3390/met12081363 83

Mengqi Yang, Chong Gao, Jianchao Pang, Shouxin Li, Dejiang Hu and Xiaowu Li et al.
High-Cycle Fatigue Behavior and Fatigue Strength Prediction of Differently Heat-Treated 35CrMo Steels
Reprinted from: *Metals* 2022, 12, 688, doi:10.3390/met12040688 99

Zhenxian Zhang, Zhongwen Li, Han Wu and Chengqi Sun
Size and Shape Effects on Fatigue Behavior of G20Mn5QT Steel from Axle Box Bodies in High-Speed Trains
Reprinted from: *Metals* 2022, 12, 652, doi:10.3390/met12040652 113

Zenan Yang, Lu Liu, Jianbin Wang, Junjie Xu, Wanrong Zhao and Liyuan Zhou et al.
Revealing the Formation of Recast Layer around the Film Cooling Hole in Superalloys Fabricated Using Electrical Discharge Machining
Reprinted from: *Metals* 2023, 13, 695, doi:10.3390/met13040695 125

Robert Fleishel, William Ferrell and Stephanie TerMaath
Fatigue-Damage Initiation at Process Introduced Internal Defects in Electron-Beam-Melted Ti-6Al-4V
Reprinted from: *Metals* **2023**, *13*, 350, doi:10.3390/met13020350 . . . . **135**

Ruixue Liu, Jie Wang, Leyun Wang, Xiaoqin Zeng and Zhaohui Jin
Cluster Hardening Effects on Twinning in Mg-Zn-Ca Alloys
Reprinted from: *Metals* **2022**, *12*, 693, doi:10.3390/met12040693 . . . . **151**

Preface to "Deformation, Fracture and Microstructure of Metallic Materials"

Exploring the relationship between deformation, fracture, and the microstructure of metallic materials is vital for their development. Thus, the works collected in this reprint investigate the microstructure-related deformation and fracture behavior of steels, superalloys, titanium alloys, etc. Sánchez et al. and Li et al. have summarized the technology and method for exploring the fracture microstructure and improving the mechanical properties of metallic materials. Qiu and Inoue, Cao et al., Du et al., Li et al., Yang et al., and Zhang et al. have investigated the microstructure-property relationship of steel materials. Yang et al., Fleishel et al., and Liu et al. have offered in-depth understanding and analysis of other metals and alloys. These works are helpful for the researchers who conduct research on the development of high-performance metallic materials and are also beneficial for promoting the application process. I would like to thank all the authors for their contributions and the managing office of Metals (MDPI) for their support in the development of this reprint.

Xiao-Wu Li and Peng Chen
Editors

Editorial
Deformation, Fracture and Microstructure of Metallic Materials

Xiaowu Li * and Peng Chen *

Key Laboratory for Anisotropy and Texture of Materials, Ministry of Education, Department of Materials Physics and Chemistry, School of Materials Science and Engineering, Northeastern University, Shenyang 110819, China
* Correspondence: xwli@mail.neu.edu.cn (X.L.); chenpeng@mail.neu.edu.cn (P.C.)

1. Introduction

Metallic materials are mostly a combination of metallic elements, such as iron, aluminum, magnesium, titanium and manganese, which may also include small amounts of non-metallic elements, such as carbon, nitrogen and oxygen. As the first materials that humans discovered and applied, metallic materials not only play significant roles in human civilization but have also been widely and irreplaceably used in modern engineering. This is due to the fact that their great potential in terms of properties, quantities and qualities could evolve and renew with growing demand, and their remarkably complex properties are adaptable to the requirements of daily life and technology. Therefore, it is necessary to endlessly seek new metallic materials that have outstanding mechanical properties or modify existing materials to improve their mechanical properties; in this way, an in-depth understanding of the deformation, fracture and microstructure of various metallic materials is of particular significance.

As is well known, the deformation and fracture mechanisms of materials are strongly dependent on their initial microstructures (e.g., grain size, grain boundary character, inclusion, precipitate, phase composition, microstructural and chemical nonuniformity), which play a determining role in defining their mechanical properties. In addition, understanding the evolution of a microstructure during deformation is also extremely important for understanding the deformation and fracture mechanisms.

2. Contributions

Eleven papers, including two review papers and nine research papers, have been published focusing on the microstructure-related deformation and fracture behavior of steels, superalloys, titanium alloys and so on. Subsequently, an overview of the contributions is given as follows:

The two review papers summarize the technology and methods deployed in the exploration of fractured microstructures and improving the mechanical properties of metallic materials. Sánchez et al. [1] reviewed the adoption of mini-CT specimens for the fracture characterization of ferritic steels, particularly focusing on those used in the nuclear industry. The main existing results are displayed, and the main scientific and technical issues are thoroughly discussed. Li et al. [2] summarized the recent progress in the theoretical models and mechanisms of twin-related grain boundary engineering (GBE) optimization. An appropriate GBE treatment has been confirmed to be an effective pathway to improve some special mechanical properties (e.g., high-temperature tension, creep, low-cycle fatigue, etc.); thus, it is a feasible method for the development of high-performance metallic materials.

Steel materials are the most important metallic materials and have a wide range of applications, and the relevant investigations on the microstructure—property relationship are of the greatest importance. Qiu and Inoue [3] traced the evolution of Poisson's ratio during the tensile deformation process in a low-carbon hot-rolled steel. The results revealed that the average Poisson's ratio could not accurately express the local Poisson's ratio in the discontinuous-yielding regime, and the Poisson's ratio varied significantly within a plastic band in this phase. Cao et al. [4] designed a special-shaped epoxy steel sleeve (SSESS) to repair low-strength tee pipes, and the reliability of the repairs was proven in repairing tests, hydraulic burst tests, and simulations. In Du et al. [5], the effect of strain rate on the tension deformation behavior of an Fe-30Mn-8Al-1.0C low-density austenitic steel was elucidated. Their results indicated that a good strength–ductility combination was achieved in the sample deformed at 10^{-3} s^{-1}; in this case, microbands and deformation twins were observed. Thus, the combination of microband-induced plasticity together with twinning-induced plasticity (TWIP) leads to a continuous strain hardening behavior and, consequently, to superior mechanical properties. Li et al. [6] investigated the quenching residual stress of hot-rolled seamless steel tubes with different cooling intensities by using ANSYS simulation software. Their results offered data support and theoretical reference for the heat treatment process design of seamless steel tubes. Yang et al. [7] have explored the microstructure, tensile properties, fatigue properties, and fatigue cracking mechanisms of 35CrMo steel processed by four heat-treatment procedures to obtain optimum fatigue performance. In addition, a suitable formula for fatigue strength prediction of the Cr–Mo steel was established on the basis of corresponding fracture mechanisms. Zhang et al. [8] conducted the axial loading fatigue tests on the G20Mn5QT steel applied in axle box bodies of high-speed trains and studied its size and shape effects on fatigue behavior. Their work is beneficial for the design of axle box bodies in high-speed trains.

There are also many other metals and alloys (e.g., superalloys, Ti alloys, and Mg alloys) used in different applications, and the corresponding relationship between microstructures and properties also needs to be comprehensively understood. In Yang et al. [9], high-resolution transmission electron microscopy (TEM) was used to study the recast layer formed by electrical discharge machining a single-crystal superalloy, considering the significant role of the microstructure of the recast layer for the performance of single-crystal blades. The results showed that the recast layer is in the condition of a supersaturated solution with a single-crystal structure epitaxially grown from the matrix. Fleishel et al. [10] intentionally induced the defects generated by the electron beam melting (EBM) of Ti-6Al-4V and investigated their influences on fatigue life. The reduced fatigue life caused by these defects and the relation of defect morphology to the material failure were investigated. Liu et al. [11] performed molecular dynamics (MD) simulations to study the interaction between Zn-Ca clusters and twin boundaries (TBs) to clarify the pronounced hardening effects, which is favorable to understanding the work hardening behavior of Mg-Zn-Ca alloys.

3. Conclusions and Outlook

This Special Issue aims to collect the latest scientific achievements in the microstructure-related deformation and fracture behavior of various metallic materials under monotonical or cyclic loads. According to the research findings arising from this collection of works, the initial microstructure and microstructural evolution have a significant effect on deformation and fracture mechanisms and, thus, mechanical properties. To this end, various microstructure characterizations, mechanical tests and numerical simulations have contributed to this Special Issue. These results are beneficial for promoting the potential applications of the involved materials and for the future development of novel high-performance materials. Finally, I would like to thank all of the authors for their contribution and the managing office of *Metals* (MDPI) for their support in the development of this Special Issue.

Conflicts of Interest: The authors declare no conflict of interest.

References

1. Sánchez, M.; Cicero, S.; Kirk, M.; Altstadt, E.; Server, W.; Yamamoto, M. Using Mini-CT Specimens for the Fracture Characterization of Ferritic Steels within the Ductile to Brittle Transition Range: A Review. *Metals* **2023**, *13*, 176. [CrossRef]
2. Li, X.; Guan, X.; Jia, Z.; Chen, P.; Fan, C.; Shi, F. Twin-Related Grain Boundary Engineering and Its Influence on Mechanical Properties of Face-Centered Cubic Metals: A Review. *Metals* **2023**, *13*, 155. [CrossRef]

3. Qiu, H.; Inoue, T. Evolution of Poisson's Ratio in the Tension Process of Low-Carbon Hot-Rolled Steel with Discontinuous Yielding. *Metals* **2023**, *13*, 562. [CrossRef]
4. Cao, J.; Jia, H.; Ma, W.; Wang, K.; Yao, T.; Ren, J.; Nie, H.; Liang, X.; Dang, W. Repair Reliability Analysis of a Special-Shaped Epoxy Steel Sleeve for Low-Strength Tee Pipes. *Metals* **2022**, *12*, 2149. [CrossRef]
5. Du, J.; Chen, P.; Guan, X.; Cai, J.; Peng, Q.; Lin, C.; Li, X. The Effect of Strain Rate on the Deformation Behavior of Fe-30Mn-8Al-1.0C Austenitic Low-Density Steel. *Metals* **2022**, *12*, 1374. [CrossRef]
6. Li, Z.; Zhang, R.; Chen, D.; Xie, Q.; Kang, J.; Yuan, G.; Wang, G. Quenching Stress of Hot-Rolled Seamless Steel Tubes under Different Cooling Intensities Based on Simulation. *Metals* **2022**, *12*, 1363. [CrossRef]
7. Yang, M.; Gao, C.; Pang, J.; Li, S.; Hu, D.; Li, X.; Zhang, Z. High-Cycle Fatigue Behavior and Fatigue Strength Prediction of Differently Heat-Treated 35CrMo Steels. *Metals* **2022**, *12*, 688. [CrossRef]
8. Zhang, Z.; Li, Z.; Wu, H.; Sun, C. Size and Shape Effects on Fatigue Behavior of G20Mn5QT Steel from Axle Box Bodies in High-Speed Trains. *Metals* **2022**, *12*, 652. [CrossRef]
9. Yang, Z.; Liu, L.; Wang, J.; Xu, J.; Zhao, W.; Zhou, L.; He, F.; Wang, Z. Revealing the Formation of Recast Layer around the Film Cooling Hole in Superalloys Fabricated Using Electrical Discharge Machining. *Metals* **2023**, *13*, 695. [CrossRef]
10. Fleishel, R.; Ferrell, W.; TerMaath, S. Fatigue-Damage Initiation at Process Introduced Internal Defects in Electron-Beam-Melted Ti-6Al-4V. *Metals* **2023**, *13*, 350. [CrossRef]
11. Liu, R.; Wang, J.; Wang, L.; Zeng, X.; Jin, Z. Cluster Hardening Effects on Twinning in Mg-Zn-Ca Alloys. *Metals* **2022**, *12*, 693. [CrossRef]

Disclaimer/Publisher's Note: The statements, opinions and data contained in all publications are solely those of the individual author(s) and contributor(s) and not of MDPI and/or the editor(s). MDPI and/or the editor(s) disclaim responsibility for any injury to people or property resulting from any ideas, methods, instructions or products referred to in the content.

Review

Using Mini-CT Specimens for the Fracture Characterization of Ferritic Steels within the Ductile to Brittle Transition Range: A Review

Marcos Sánchez [1,*], Sergio Cicero [1,*], Mark Kirk [2], Eberhard Altstadt [3], William Server [4] and Masato Yamamoto [5]

1 Laboratory of Materials Science and Engineering (LADICIM), University of Cantabria, E.T.S. de Ingenieros de Caminos, Canales y Puertos, Av./Los Castros 44, 39005 Santander, Spain
2 Phoenix Engineering Associates, Inc., 119 Glidden Hill Road, Unity, NH 03743, USA
3 Helmholtz-Zentrum Dresden–Rossendorf, Bautzner Landstrasse 400, 01328 Dresden, Germany
4 ATI Consulting, Black Mountain, NC 28711, USA
5 Central Research Institute of Electric Power Industry, Nagasaka 2-6-1, Yokosuka 240-0196, Japan
* Correspondence: sanchezmam@unican.es (M.S.); ciceros@unican.es (S.C.)

Abstract: The use of mini-CT specimens for the fracture characterization of structural steels is currently a topic of great interest from both scientific and technical points of view, mainly driven by the needs and requirements of the nuclear industry. In fact, the long-term operation of nuclear plants requires accurate characterization of the reactor pressure vessel materials and evaluation of the embrittlement caused by neutron irradiation without applying excessive conservatism. However, the amount of material placed inside the surveillance capsules used to characterize the resulting degradation is generally small. Consequently, in order to increase the reliability of fracture toughness measurements and reduce the volume of material needed for the tests, it is necessary to develop innovative characterization techniques, among which the use of mini-CT specimens stands out. In this context, this paper provides a review of the use of mini-CT specimens for the fracture characterization of ferritic steels, with particular emphasis on those used by the nuclear industry. The main results obtained so far, revealing the potential of this technique, together with the main scientific and technical issues will be thoroughly discussed. Recommendations for several key topics for future research are also provided.

Keywords: mini-CT; ductile-to-brittle transition range; reference temperature; master curve

1. Introduction

Reactor pressure vessels (RPVs) are safety-critical components in nuclear power plants (NPPs). To ensure the continued operation of NPPs the fracture resistance of the RPV beltline materials is monitored throughout the plant's lifetime. RPVs are made of ferritic steels, which fail via a ductile mechanism at relatively high temperatures but transition to brittle fracture at lower temperatures. Additionally, the transition from ductile to brittle behavior is shifted towards higher temperatures when these steels are exposed to neutron irradiation. To ensure that ferritic steels maintain adequate structural integrity at service temperatures, actual RPV materials are included in surveillance programs that evaluate toughness behavior during their service life. These surveillance programs were originally based on impact energy measured from Charpy specimens. However, Charpy testing is used in a semi-empirical approach that cannot directly measure the material's fracture toughness. In the past several decades, a direct evaluation of the fracture behavior of RPV steels within the ductile to brittle transition range (DBTR) has been enabled by the master curve (MC) methodology, which has gained increased acceptance in recent years.

The MC is an engineering approach that provides a means to characterize the fracture behavior of ferritic steels within the DBTR [1,2]. The MC is standardized by ASTM E1921 [3]

and by JEAC4216 [4]. It is based on the weakest link theory and, thus, describes the fracture behavior using a three-parameter Weibull distribution. Two of the parameters, the location parameter (K_{min}) and the shape parameter (b), have been empirically defined for all ferritic steels (taking values of 20 MPa·m$^{0.5}$ and 4, respectively), whereas the scale parameter (K_0) has also been defined in terms of the material reference temperature (T_0). Thus, testing is performed to estimate this single material parameter. T_0 represents the temperature at which the median of fracture toughness, K_{Jcmed}, for a 1T (meaning 1-inch, or 25.4 mm) thick specimen is equal to 100 MPa·m$^{0.5}$. Once T_0 is estimated from K_{Jc} data for the material being analyzed, the MC can be defined for any probability of failure (P_f) by the following Equation (1):

$$K_{JC,\,Pf} = 20 + \left[\ln\left(\frac{1}{1-P_f}\right)\right]^{1/4} \cdot \{11 + 77 \cdot \exp[0.019 \cdot (T - T_0)]\} \quad (1)$$

In principle, T_0 can be defined by testing K_{Jc} specimens of any thickness. These test data are then scaled to a "1T-equivalent" K_{Jc} value, 1T". Thus, for miniature compact tension (mini-CT) specimens, the MC can be used to convert the measured K_{Jc} value into the corresponding $K_{Jc(1T)}$ equivalent, using the following equation (B being the thickness of the tested specimen, which is 4 mm for the mini-CT (2):

$$K_{JC(1T)} = 20 + [K_{JC} - 20]\left(\frac{B}{25.4}\right)^{1/4} \quad (2)$$

Hence, the MC addresses the three main characteristics of fracture toughness characterization within the DBTR: the scatter of the results, the dependency of fracture toughness on temperature, and the adjustment for test specimen thickness. Figure 1 shows an example of MC obtained by the authors in an ongoing program [5,6]. It is noted that the orange line defines a limiting condition in MC to consider the effect of the plastic zone evolution ahead of the crack tip, as discussed later in Section 2.5. All the K_{Jc} data above this line (red symbols) will be censored to account appropriately for their effect on T_0.

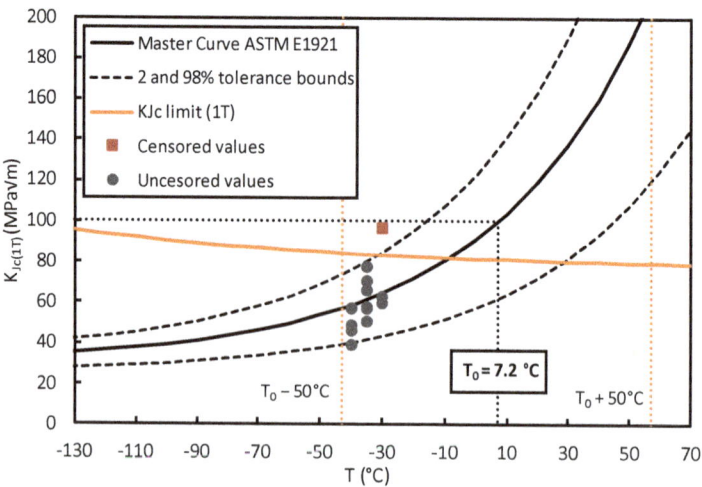

Figure 1. MC for A533B LUS, obtained using mini-CT specimens. Censoring criterion as defined by ASTM E1921 [3].

The need for accurate characterization of the DBTR faces the problem that the availability of material for fracture testing is sometimes limited. However, NPPs generally have a large number of irradiated and previously tested Charpy specimens, so the possibility of performing further testing with this remnant material is of great practical interest. Such testing can be performed using mini-CT specimens, knowing that one tested Charpy specimen allows the fabrication of a maximum of eight 4 mm-thick mini-CT specimens. This mini-CT testing approach brings several benefits, such as: (a) the direct assessment of fracture toughness rather than the semi-empirical approach based on Charpy measurements; (b) the ability to characterize the local properties of heterogeneous materials; (c) a significant increase in the surveillance monitoring database providing greater confidence in the data; (d) a reduction in the volume of irradiated material needed for characterization; and (e) the possibility for re-orientation of the notch in the base material (e.g., T-L vs. L-T) becomes possible, which is particularly important for older plants that have only L-T orientation data while the current ASME Code uses T-L orientation data.

The purpose of this review is to collect and summarize the available scientific and technical information about testing mini-CT fracture specimens, contribute to the development of this miniaturization technique, and provide insights about the main remaining challenges for testing, evaluation, and standardization efforts. A generic issue with the use of mini-CT specimens is associated with the small size itself. The stress intensity factor is a function of the far-field load and the absolute crack size (i.e., $K_I \sim \sigma_0 \sqrt{a}$). Therefore, at a given fracture mechanic load K_I, the relative size of the plastic zone (plastic volume/specimen volume) is larger in mini-CTs as compared to larger specimens. Consequently, the violation of the small-scale yielding criterion (and thereby the loss of constraint) starts at lower K-values in mini-CTs, reducing the measuring capacity of mini-CT specimens compared to larger specimens. The implications of this for MC testing are addressed in Section 2.5.

2. Experimental Challenges Presented by the Mini-CT

2.1. The Geometry of Mini-CT Specimens

CT specimens are one of the most common types of standardized specimens used in fracture mechanics testing. The geometry provides an efficient use of the tested material, with the majority of the sample volume used to establish a controlled stress state at the crack tip during loading. However, the miniaturization of CT specimens entails a series of specific testing challenges that are discussed next. This review focuses on 4 mm-thick CT specimens, which may be found in the literature under different names, the most common being mini-CT (the one used in this document), 0.16T-CT, or MCT specimens.

Principally, two different mini-CT geometries have been proposed in the literature, although some minor modifications may be found and will be mentioned in this document where necessary. Figure 2 shows both geometries, also showing a comparison between mini-CT specimens and larger CT specimens (Figure 2b). The first one is the reduced normalized geometry for the CT specimen given by the ASTM E1921 [3] standard: its dimensions are $10 \times 9.6 \times 4$ mm^3 (e.g., [7]). Here, it is important to note that the 2021 version of the standard ASTM E1921 [3] permits the use of this mini-CT geometry for the MC characterization. The second geometry is designed to capture directly the geometry of the Charpy specimen: its dimensions are $10 \times 10 \times 4.2$ mm^3 (the thickness actually ranges between 4 and 4.2 mm) (e.g., [8]). The latter has the advantage of simplified machining, which may be especially important with irradiated materials, although it does not accurately reflect the geometry established by the ASTM E1921 standard [3]. Nevertheless, the literature suggests that this geometric discrepancy does not significantly affect the fracture toughness results, given that the differences in the crack tip stress conditions are small [8,9]. Although it is not clear which geometry will be predominant in the future, in the present review, the specimen that strictly complies with the requirements of the standard ASTM E1921 [3] ($10 \times 9.6 \times 4$ mm^3) will be referred to as (standardized) mini-CT, while the $10 \times 10 \times 4.2$ mm^3 geometry will be referred to as modified mini-CT. Figure 2b shows different CT specimen sizes, including mini-CTs.

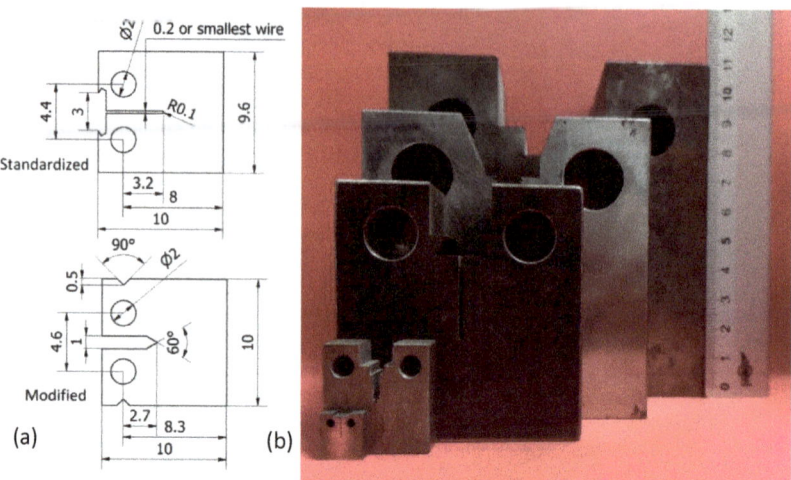

Figure 2. (a) Standardized mini-CT vs. modified mini-CT (dimensions in mm); (b) image showing the scales of different CT specimens and a standardized mini-CT.

Regarding the standardized mini-CT specimen, ASTM E1921 [3] permits three different CT specimen configurations. However, in practice, only the one without a cut-out section (derived from ASTM E399 [10]) has been employed for mini-CTs due to the difficulty of mounting a clip gauge inside the mini-CT specimen to directly measure the load-line displacement; the purpose of the other two specimens with cut-out sections (see Section 2.2 for further discussion) is to allow room for a clip gauge to measure load-line displacement. One advantage of ASTM E1921 [3] is that it does not establish a specific specimen size, but rather all the dimensions are set relative to the specimen thickness, B. Here, it is important to note that, following [3], the required tolerance of all the dimensions is ±0.013 W, corresponding to ±0.1 mm for mini-CT specimens, and the maximum clearance between the pin and the hole is 0.02 W (0.16 mm for mini-CT specimens). Finally, the maximum allowable starter notch dimension shall not exceed 0.063 W, which is about 0.5 mm for mini-CT specimens.

2.2. Load Line vs. Front Face Displacement

As mentioned above, an important topic that affects the geometry of the mini-CT specimen is related to the location at which displacement is measured during fracture tests. Clip gauges may be placed either on the load line (LL) or at the front face (FF) position. Given that mounting a clip gauge inside mini-CTs is generally not feasible, Scibetta et al. [11] proposed a method to measure the LL displacement outside of the mini-CT specimen using a dedicated clip gauge. Its main contribution is related to the use of notches on the top and bottom surfaces of the specimen to allow a clip gauge with sharp razor blades to be placed at the LL position. This technique was also employed by Sokolov [12], who determined that external clip gauges improved the reliability and sensitivity of the measurements when compared to those derived from integrated front-face cut-off notches, such as those suggested in [13–15]. The literature distinguishes two main advantages of using this external LL measurement location: (a) the simplicity of handling such a small specimen and clip gauge in the hot cell or other remote conditions, and (b) more rigidity of the specimen in the region of the loading pins. As a disadvantage, the resulting measurements could be affected when plastic deformation occurs in the vicinity of the pins [8]. Chaouadi et al. [8] compared the displacement measurements on the FF (v_{FF}) versus the LL (v_{LL}). The load-displacement test records of two specimens with medium and high toughness were compared, showing that they lead to comparable K_{Jc} values. Besides, the ratio between the two measured displacements was about 0.72, in close agreement with those values suggested in the literature (e.g., [16,17]).

As for the FF displacement, ASTM E1921 [3] allows this measurement technique by applying a conversion factor (R) of 0.73, as an alternative to measuring the LL displacement. This conversion factor was derived by Landes [16] from conventional CT specimens by using Equation (3):

$$R = \frac{v_{LL}}{v_{FF}} = \frac{a/W + r \cdot (1 - a/W)}{a/W + r \cdot (1 - a/W) + X/W} \quad (3)$$

where r is the ratio of the distance between the crack tip and the rotation center to the ligament size, and X is the offset between the front face and load line. Landes recommended the value of r = 0.33 and demonstrated that the sensitivity of R on r (dR/dr) is moderate. The conversion factor of R = 0.73 for a standard CT specimen with a/W = 0.5 and X/W = 0.25 corresponds to r = 0.352. In practice, the pre-cracking procedure will not always result in the ideal value of a/W = 0.5; the allowable range is 0.45 to 0.55 [3]. With r = 0.352, this corresponds to a conversion factor range of R = 0.72 to 0.74. In case of deviations from the standard geometry with X/W > 0.25 (e. g. for better applicability of extensometers at FF), the conversion factor R can be significantly affected (e.g., X/W = 0.375 → R = 0.64). Therefore, the evaluation of R with Equation (3) is preferable to a constant value of R = 0.73.

Miura et al. [17] investigated both the conversion factor and the rotation center location factor for mini-CT specimens. Finite element analyses were performed considering three material models with different plastic hardening behaviors. The conversion factor converged within the range of 0.73 to 0.75, and the point of rotation was located at the center of the ligament during loading. These analytic results were then examined using an experimental dataset of mini-CT specimens. The effect of selecting the conversion factor either as 0.73 or 0.75 had a minor impact on the evaluation of the fracture toughness and the estimated T_0 value.

2.3. Side Grooving

In fracture mechanics tests, the specimens are often side grooved to ensure that the crack front after precracking meets the straightness criteria of the testing standards and to improve the uniformity of the stress state along the crack front. However, this technique was found to have little effect on mini-CT specimens [8,18,19] as long as testing is performed within the DBTR. Wallin et al. [19] demonstrated that side grooving has a minor effect on the location of cleavage initiation, although further experimental evidence is warranted. On the other hand, side grooving inherently reduces the measurement capability of the specimen, which is already low in miniature specimens. Yamamoto et al. [18] analyzed the statistical distribution of the data sets with and without side grooving, resulting in a Weibull modulus of 5.1 and 5.5, respectively, thus concluding that side grooving does not affect significantly the statistical distribution of fracture toughness.

2.4. Crack Front Curvature

One challenge when using miniature specimens is generating a sufficiently straight crack front during fatigue pre-cracking. The stress state along the crack front tends to promote fatigue crack growth at the center of the specimen, and the resulting crack front is generally parabolic. In addition, the residual stress state, variation in material properties, or machine misalignment can cause uneven crack growth, resulting in a slanted crack front. For curved crack fronts, the variations in the J-integral and the constraint conditions along the crack front can differ from those existing in straight cracks, which is the crack geometry on which the equations in ASTM E1921 [3] are based.

Lambrecht et al. [20] studied the effect of crack curvature on the T_0 results obtained using mini-CT specimens. They observed a negligible effect of crack front straightness on T_0 by comparing the T_0 values obtained with valid and invalid crack front curvatures, as defined by ASTM E1921 [3]. They also suggested discarding the outermost points from the crack front curvature assessment, something later adopted by ASTM E1921-21 [3].

Lindqvist et al. [21] investigated the effect of crack front curvature on fracture toughness within DBTR by using numerical and experimental methods. The results supported the relaxation of the curvature acceptance criterion proposed by ASTM E1921 [3]. For the investigated crack front curvatures, the effect of curvature on T_0 was smaller than the uncertainty of the T_0 estimations. The authors concluded that curved crack fronts tend to have slightly higher T_0 values, which is conservative.

2.5. Temperature-Related Issues

At low test temperatures, the freezing of the specimen, extensometer, and clevis may affect the experimental measurements. Ice on the clevis and the specimen hinder the connection between the specimen and the extensometer, and this sometimes leads to load-deflection measurements that do not reflect specimen behavior. Furthermore, the possible change in the electric signal of the extensometer needs to be compensated at low temperatures. This phenomenon was studied in an interlaboratory study [14], where tests were carried out with a clip gauge set in the front face of the specimens. It was found that at $-150\ °C$, the clip gauges tended to provide about 3 to 7% larger readings than the actual values due to the change in the electric resistance of the extensometer. Since the linearity between the signal change and the deflection change was maintained, a temperature-dependent coefficient was used for converting the electric signal into the deflection measurement at low temperatures.

On the other hand, the test temperature has to be monitored and controlled on the specimen surface, as established by ASTM E1921 [3]. However, some authors found difficulties in welding a thermocouple on the surface due to specimen size restrictions, and additionally, the effect of heat input on fracture toughness is not clear for such small specimens. Therefore, several authors (e.g., [7,18]) decided to control the test temperature by means of a thermocouple welded on the surface of the clevis. The temperature difference between the specimen and the clevis was found negligible after holding the specified temperature for at least 15 min [7]. However, other work [22] decided to monitor directly the test temperature with a thermocouple attached to the surface of the specimen.

Another important topic when dealing with the mini-CT specimen is determining the number of specimens required to obtain a valid T_0 value, which is intrinsically related to the temperature range selected for the tests. It is necessary to mention the limitations on testing conditions imposed directly by the ASTM E1921 standard [3]. Two of the most important restrictions are the $T_0 \pm 50\ °C$ testing temperature range, and the $K_{Jclimit}$ value for the fracture toughness results (see orange curve in Figure 1), which is the limiting value for data censoring:

$$K_{Jclimit} = \sqrt{\frac{E \cdot b_0 \cdot \sigma_y}{30 \cdot (1 - \nu^2)}} \qquad (4)$$

where b_0 is the remaining ligament, σ_y is the material yield strength, E is the elastic modulus, and ν is the Poisson's ratio. $K_{Jclimit}$ ensures that the remaining ligament has sufficient size to ensure high constraint conditions at the crack front and that small-scale yielding conditions are met. The remaining ligament, b_0, is proportional to the thickness of the specimen, thus, as shown in Equation (4), the smaller the ligament, the smaller the $K_{Jclimit}$. Hence, small specimens such as mini-CTs have low $K_{Jclimit}$ values, which forces testing at rather low temperatures to ensure that most measured K_{Jc} values fall below $K_{Jclimit}$; this reduces the temperature range over which mini-CT tests can be conducted.

Another aspect that conditions the $K_{Jclimit}$ is the material yield strength. In this sense, Sugihara et al. [23] analyzed the effect of this material property in the $K_{Jclimit}$ of irradiated materials, and therefore its impact on the validity window of the MC approach. For irradiated conditions for the same test temperature, σ_y is higher than in unirradiated conditions. On the other hand, the change in T_0 due to irradiation embrittlement leads to higher test temperatures for irradiated materials than for unirradiated materials. For these reasons, if the decrease in σ_y by increasing the test temperature is greater than the increase in σ_y caused by irradiation, the σ_y of the irradiated material may be lower than that of the unirradiated material at the corresponding test temperatures. Therefore, $K_{Jclimit}$ may be even lower in irradiated conditions than in unirradiated conditions. The authors studied this effect by using literature data [24] finding no significant tendencies for this particular case.

Regarding the test temperature, ASTM E1921 [3] recommends the selected temperature be close to that at which the K_{Jcmed} value is approximately 100 MPa·m$^{0.5}$ for the specimen size being used. Based on that, Tobita et al. [22] proposed that for mini-CT specimens, the test temperature equivalent to 100 MPa·m$^{0.5}$ would be given by the K_{Jcmed} MC expression together with the corresponding thickness correction, as shown in Equation (5).

$$100 = \frac{\{30 + 70 \cdot \exp[0.019 \cdot (T - T_0)]\} - 20}{\left(\frac{B_{0.16T}}{B_{1T}}\right)^{0.25}} + 20 \tag{5}$$

Clearing for the equation, resulted in $T = T_0 - 29°C$. However, considering that the precision of the test temperature control is ± 3 °C, the optimum test temperature to minimize the likelihood of invalid K_{Jc} values (i.e., K_{Jc} values larger than $K_{Jclimit}$) was selected as $T = T_0 - 32°C$.

Moreover, Miura et al. [7] reported, for the unirradiated mini-CT specimens, the range of temperatures that would lead to reducing the invalid data due to $K_{Jclimit}$, improving the efficiency when obtaining valid (non-censored) data. They observed that the ratio of valid data (the number of uncensored values to the total number of tests) for the mini-CT specimens increased when T-T_0 was reduced and reached unity when T-T_0 was less than -30 °C. This trend agrees well with the reference curve from ASTM E1921 [3], which was obtained for PCCv (pre-cracked Charpy) specimens. Therefore, the recommendation was to test mini-CT specimens within the range of -50 °C $\leq T - T_0 \leq -30$ °C. This test temperature range was subsequently confirmed on materials in both irradiated and unirradiated conditions [8,23,25].

In terms of the number of specimens, the ASTM E1921 standard [3] suggests a minimum number of valid tests ranging from 6 to 9 specimens, depending on the testing temperatures, in order to determine a valid T_0. At the same time, it is well known that the uncertainty in T_0 determination increases when the lower shelf is approached, which is otherwise necessary (as shown above) to obtain a sufficient number of uncensored K_{Jc} data. This phenomenon may be countered by increasing the number of tests since the uncertainty of T_0 is inversely proportional to the square root of the number of uncensored specimens, as shown in Equation (6).

$$\sigma_{T0} = \left(\frac{\beta^2}{r'} + \sigma_{exp}^2\right)^{1/2} \tag{6}$$

where β is the sample size uncertainty factor determined following Section 10.9.1 in ASTM E1921 [3], r' is the total number of uncensored data used to calculate T_0 and σ_{exp} is the contribution of experimental uncertainties, usually taken as 4 °C.

In this regard, Chaouadi et al. [8] determined the minimum number of specimens leading to a reliable T_0 value, assuming that this will be the one calculated using the largest number of specimens. For this purpose, a set of mini-CT tests of different materials was examined, and the transition temperature was iteratively calculated as the number of tests used in the calculation increased. The analysis determined that the T_0 stabilized within ± 2 °C after using about 16 specimens in the calculation and found that the minimum number as required per ASTM E1921 [3] is mostly between 8 and 10 specimens, or even more. It should be noted that this minimum number is based on experimental data for materials with known T_0. Thus, for materials that are characterized solely by mini-CT, a larger number of specimens may be required.

Another criterion imposed by ASTM E1921 [3] is the ductile crack growth (DCG) limitation. This criterion established that DCG cannot exceed either 0.05 (W-a_0) or 1 mm, being in practice around 0.2 mm for mini-CT specimens. This is particularly important for low upper-shelf (LUS) materials. As was reported in several works [25–27], this type of material may exhibit DCG at temperatures near T_0, complicating the selection of testing temperatures. Therefore, the authors recommended testing more than 15 specimens to properly carry out an MC evaluation for LUS materials.

In summary, to obtain valid results, it is recommended to test at least 30 °C below the final T_0, increasing the chances of obtaining non-censored results (see the shaded area in Figure 3). Additionally, the testing temperature cannot be below ($T_0 - 50$ °C) according to the requirements of ASTM E1921 [3] and JEAC4216 [4], and, finally, T_0 is not known exactly beforehand. Figure 3 shows the ratio of valid data (non-censored test results to the total number of tests) versus the test temperature ($T - T_0$) in a number of experimental results. Each point represents a set of experimental results performed on a given material at a given temperature. It can be observed how the range -50 °C $\leq T - T_0 \leq -30$ °C maximizes the ratio of valid data. The figure also shows the ASTM E1921 [3] requirement of valid data, which depends on the test temperature and the resulting T_0, revealing how it fits the experimental results. Finally, the red line in Figure 3 shows a proposal for a hyperbolic tangent best-fit curve (non-linear least squares):

$$f(x) = A + B \times \tanh((x - C)/D) \tag{7}$$

where A, B, C, and D are the four parameters required for the adjustment of the curve, which takes values of 0.802, -0.196, -17.16, and 11.06, respectively. The curve reveals how the ratio of valid data is close to 1 as long as $T - T_0 \leq -30$ °C, in agreement with the results shown above (e.g., [7,8,23,25]).

Recently, efforts have been made to address these issues. Yamamoto et al. [28] published a paper that aims to extend the validity temperature range below ($T_0 - 50$ °C) by defining new criteria that allow the inclusion of data in the MC evaluation that otherwise would be rejected.

To conclude with the temperature-related issues, it is clear from the information summarized here that knowing an initial estimation of T_0 is beneficial, given that the above discussions assumed that the T_0 value was already known for the studied material. In many situations (e.g., when surveillance data are available), T_0 can be estimated from existing Charpy data (e.g., [3,23,25]), but this will not be possible in all situations. When Charpy impact transition curves are available, the ASTM E1921 [3] standard provides a procedure for selecting a test temperature (T) in the neighborhood of T_0. Indeed, the ASTM E1921 [3] standard proposes the following relationship between test temperature, T, and T_{41J}:

$$T = T_{41J} + C \tag{8}$$

where the constant C is given for various specimen thickness values that can be fitted with the following equation:

$$C = 14.845 \times \ln[\text{thickness(mm)}] - 71.8 \text{ ; °C} \tag{9}$$

For the mini-CT specimen, this equation provides $T = T_{41J} - 51$ °C [7]. However, this estimate does not guarantee that the resulting test temperature is appropriate for mini-CT specimens.

Equation (8) is based on the work by Sokolov and Nanstad [24], who studied the relationship between T_0 and T_{41J}. The authors derived the following well-known expression:

$$T_0 = T_{41J} - 24 \text{ °C} \tag{10}$$

The general trend, according to the reviewed literature, is that the Sokolov correlation [24] can be used, as shown in Figure 4.

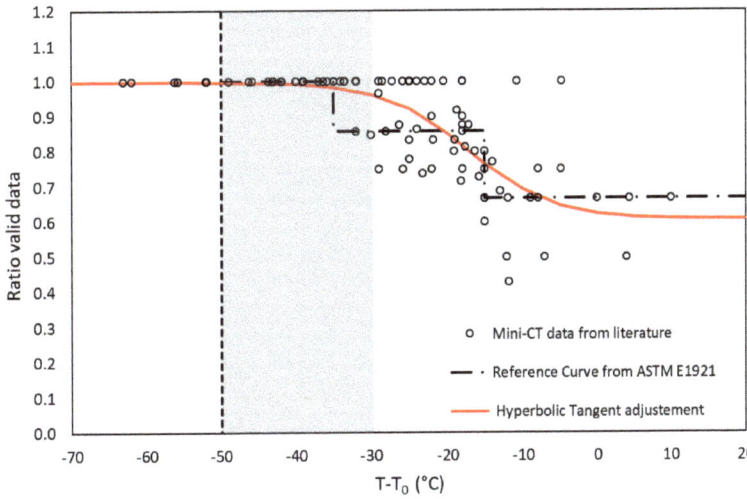

Figure 3. Relation between the ratio of valid data vs. (T − T$_0$) for mini-C(T) specimen. T is the testing temperature. Data from: [7–9,12–14,22,23,26,29–34].

As was mentioned above, Tobita et al. [22] intended to define an initial test temperature by providing a K$_{Jcmed}$ value of 100 MPa·m$^{0.5}$ by the following equation: T = T$_0$ − 32 °C. Thus, considering Equation (10), they suggested the following equation to select the test temperature for mini-CT specimens:

$$T = T_0 - 32\,°C = T_{41J} - 56\,°C \tag{11}$$

Further refinement of Tobita's approach has been documented in subsequent publications by JAEA [30,32,35].

More recently, Yamamoto et al. have demonstrated that even with an uncertain estimation of the initial test temperature, which is possible due to the uncertainty of Equation (10), test temperature selection for a few subsequent specimens may help to quickly recover from the issue [36]. They demonstrated that even if the initial test temperature is less than optimal due to an uncertain pre-estimated T$_0$, which was defined as up to a maximum deviation of 40 °C from the true T$_0$, the resultant success rate of T$_0$ evaluation using N = 12 or fewer specimens is quite high (98% or more) (see Table 1) if their proposed test temperature selection procedure is used.

Table 1. Influence of error in initial T$_0$ guess on the success rate of valid T$_0$ evaluation with N=12 or fewer specimens [36].

$\sigma_{YS(RT)}$, MPa	True T$_0$, °C	T$_0$ Guess − True T$_0$, °C				
		−40	−20	0	20	40
400	−50	97.9%	98.6%	99.1%	98.6%	98.5%
500	0	99.4%	98.6%	99.8%	98.9%	99.5%
600	50	100.0%	100.0%	99.9%	99.6%	99.9%

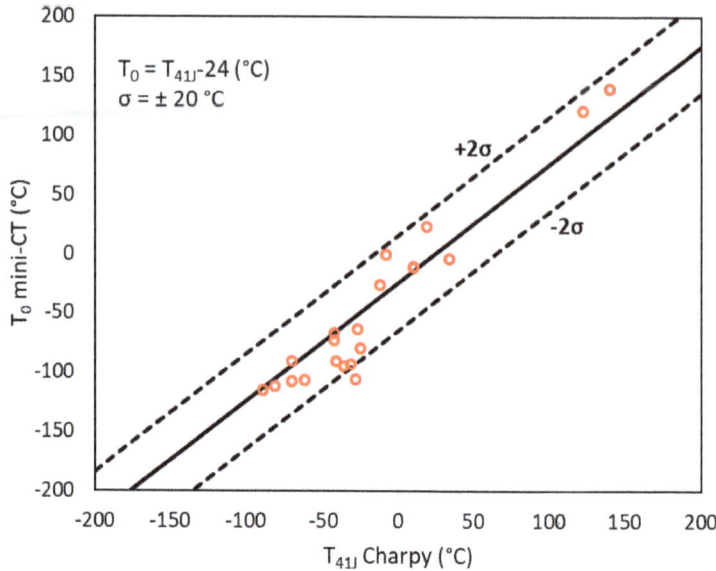

Figure 4. Correlation between mini-CT T_0 values and T_{41J} values obtained from Charpy testing. Data were taken from [7,8,11,14,22,23,30,33,35,37,38].

2.6. Loading Rate during Fracture Testing

Regarding the loading rate during the fracture testing, ASTM E1921 [3] recommends a dK/dt range of 0.1 to 2.0 MPa $\sqrt{m}$/s. Hall and Yoon [39] evaluated the loading rate dependency of T_0 values determined from PCCv specimens and other larger CT specimens for eight different materials, using the following equation:

$$T_{0R2} = T_{0R1} + B \cdot \ln(R2/R1) \tag{12}$$

where R1 and R2 are the loading rates and B, varied from 2.2 to 5.7, depending on the material. This loading rate dependency was studied in [22] for mini-CT specimens, revealing a similar behavior to that observed in larger specimens. The higher loading rates generated slightly higher T_0 values, and this tendency was almost the same for the larger specimens. In [15], it was shown that due to the proportionally larger plastic deformation developed in mini-CT specimens during fracture tests, the loading rate change in mini-CT specimens is, therefore, larger than that developed in larger specimens. However, in spite of this observation, no specific dependency of T_0 with the loading rate could be established for mini-CTs as long as the loading rate meets the ASTM E1921 [3] validity range (0.1 < dK/dt < 2 MPa·m$^{0.5}$/s).

3. Results on Unirradiated and Irradiated Steels

The DBTR of ferritic steels has been widely analyzed over the years. The main engineering tool developed to characterize the material fracture toughness within the DBTR is the MC approach [1,2], which is standardized in ASTM E1921 [3] and widely described elsewhere (e.g., [1,2]).

One of the first works related to mini-CT specimens was reported by Miura et al. [7]. They focused on two typical Japanese RPV steels, SFVQ1A forging, and SQV2A plates of two different heats. They carried out at least 6 to 8 tests per temperature, thus allowing a comparison between the single-temperature method and the multi-temperature method [3]. No significant effect was found in the T_0 evaluation method, the maximum difference being

4.7 °C. A comparison among T_0 values obtained from different CT sizes (4T, 2T, 1T, 0.4T, and 0.16T) showed a maximum deviation from the average of 4.8 °C, 4.5 °C, and 10.5 °C for SFVQ1A, SQV2A Heat 1, and SQV2A Heat 2 materials, respectively. Therefore, they concluded that T_0 can be accurately determined using mini-CT specimens.

A three-year round-robin was organized by the Central Research Institute of Electric Power Industry (CRIEPI) in Japan to verify the reliability and robustness of obtaining T_0 by means of mini-CT specimens. The first and second round robin tests [13,15] were performed on the SFVQ1A steel. The average value of the ten T_0 evaluations carried out within the first two years of the program was −102 °C, which agrees well with the T_0 values obtained with larger specimens (−91 °C to −103 °C). Besides, it was very close to the T_0 value obtained by Miura et al. [7], which, when applying the multi-temperature evaluation, resulted in a T_0 of −101 °C. Thus, the first and second round robin tests [13,15] suggested that the T_0 evaluation technique using mini-CT specimens is fairly robust.

The final stage of the mini-CT round-robin program [14] carried out on unirradiated SQV2A steel showed that six laboratories were able to determine T_0 within the expected scatter range. The fracture tests were carried out as blind tests, which means that detailed material information, such as the type of material, estimated T_0, and existing fracture toughness data for the material, were not provided to the laboratories beforehand. Participants independently selected the test temperature based on the full Charpy curve of the tested material. Although the test temperatures selected by the various laboratories varied from −120 to −150 °C, the obtained T_0 values with mini-CTs were reasonably consistent with each other; the maximum difference among participants being 16 °C. The T_0 determined using all of the mini-CT test results was equal to −115 °C. This value is in good agreement with the T_0 calculated by Miura et al. [7].

Tobita et al. [22] performed an extensive experimental campaign with five types of SA533B C1.1 RPV steels with different ductile to brittle transition temperatures. All data sets gave valid T_0 values, showing a good relationship between the T_0 obtained from mini-CT specimens and those determined from 1T-CT specimens; the maximum deviation between the 1T and mini-CT T_0 values was 13 °C for steel JRM. This database was posteriorly expanded by Takamizawa et al.'s [37] work and found a difference between mini-CT results and 1T-CT results not exceeding 10–15 °C.

The applicability of mini-CT samples with the MC approach has been confirmed for several Japanese base metal RPV steels. However, it was necessary to extend and ensure the applicability of this methodology to weld metals, which is of great interest to guarantee the safety of NPPs during long-term operation (LTO). Generally, RPV weld metals are recognized as homogeneous materials, but weld metals often exhibit inhomogeneity in multi-pass bead welding. Likewise, the HAZ in the base metal beside the weld fusion line is also one of the materials to be investigated in the surveillance programs. Since the available volume of the HAZ region is limited, the utilization of small specimens is important [29]. In this sense, several experimental campaigns have been completed in the last decade. Yamamoto et al. [18] completed fracture tests on SQV2A weld metal. The T_0 values of this weld metal obtained with mini-CT and 0.5T-CT specimens demonstrated excellent agreement, both generating a T_0 value of −77 °C. Additionally, a set of mini-CT specimens were tested with side grooves, resulting in a T_0 of −80 °C. The fixed value of C = 0.019 in the MC (see Equation (1)), which defines the shape of the K_{Jcmed} curve, was evaluated as a variable. The parameter C was calculated as 0.018 and 0.021 for the weld metal and the base metal, respectively, demonstrating that both the weld metal and the base metal may be evaluated following the MC recommendation. Subsequently, this SQV2A weld metal was investigated in an international round-robin program composed of four participants [40]. All of them obtained a valid T_0 value, with a maximum difference among them of 14 °C, which was reasonably small in comparison with the previous round-robin [14].

Yamamoto et al. [29] prepared an experimental campaign with two plates of SQV2A steel and different sulfur contents (denoted as Low S and Mid S) that were joined to each other with SQV2A weld metal by the submerged arc welding method. In total, five data sets were obtained with mini-CT specimens, two from the base materials, one from the weld metal, and two from the corresponding heat-affected zones. All data sets provided a valid T_0 value. However, when comparing the results with the T_0 values determined from 0.5T-CT specimens, the difference between the mini-CTs and the 0.5T-CTs observed in HAZ materials was larger than in the other materials. A difference of 25 °C and 10 °C was found in Low S HAZ and Mid S HAZ, respectively. Micrography analysis of the HAZ specimens (6 mini-CT and 6 0.5T-CT specimens) showed that the width of the HAZ region varied from 2.7 mm to 3.3 mm among specimens. Here it is important to note that mini-CT specimens have dimensions of 10 mm × 9.6 mm, while the 0.5T-CT specimens have dimensions of 31.25 mm × 30 mm, so it is evident that in the mini-CTs a large part of their volume is composed by the HAZ material, thus allowing a more precise characterization of the fracture of this particular zone. In any case, the authors concluded that the large variation of T_0 observed in the HAZ could be caused by the inherent heterogeneities associated with this area.

Chaouadi et al. [8] studied RPV steel 22NiMoCr37 by using the modified mini-CT specimen (10 × 10 × 4.2 mm^3) in non-irradiated conditions. In addition to the findings mentioned in the previous section, they concluded that the difference in the T T_0 obtained using conventional CTs and mini-CTs is around 12 °C, while in the case of precracked Charpy specimens (PCCv) the difference is 5 °C.

Another fundamental aspect when dealing with mini-CT specimens is the similitude of the cleavage initiation process since the MC is based on the assumption that specimen size does not affect this similitude and the thickness would only cause a statistical effect that can be easily corrected by means of the well-known Equation (2). This issue was examined by Wallin et al. [19] by comparing the location of cleavage initiation sites along the crack front for different specimen sizes and configurations, focusing the effort on mini-CT specimens. The authors determined the cumulative initiation location distribution of several datasets and found that the majority of initiations took place within 40% of the specimen center ($0.3 \leq x/B \leq 0.7$, x being the distance from one side of the specimen and B being the specimen thickness) and, in addition, in about 30% of the center of the specimen, the location initiation probability was uniform. This trend of concentrating the initiation points at the center of the specimen for mini-CT specimens was also assessed for larger (conventional) single-edge notch bend (SENB) and CT specimens. The results confirmed that the specimen size or type has a minimal effect on the distribution of the initiation location sites as long as the requirements of the ASTM E1921 [3] standard are fulfilled.

So far, the review has been focused on the applicability of mini-CT specimens in unirradiated specimens, revealing the success of this technique. Now, the results on irradiated materials are presented.

Ha et al. [30] studied a Japanese A533B class 1 steel, which was named Steel B. Mini-CT specimens were taken from the halves of irradiated PCCv specimens, which were subjected to a neutron fluence of 1.1×10^{20} n/cm^2 at 290 °C. The mini-CT specimens provided a T_0 value of −11 °C, which is slightly higher than that obtained using PCCv specimens, −24 °C. Here, it is worth mentioning that the IAEA reported a bias of around 10 °C between the CT specimen type and PCCv specimens [41].

Yamamoto [31] reported an inter-laboratory effort to evaluate European JRQ material irradiated with a fluence of 1.85×10^{19} n/cm^2 at 286 °C. All the laboratories could obtain a valid T_0 value from the given number of mini-CT and PCCv specimens in both unirradiated and irradiated conditions. The results under unirradiated conditions demonstrated a good agreement in T_0 values, −68 °C for the mini-CT specimen and −67 °C for the PCCv specimen. Despite these valid T_0 results, the JRQ material demonstrated a wide fracture toughness scatter. An excessive number of K_{Jc} values (13% for mini-CTs and 25% for PCCv specimens) were located outside the MC bounds. This trend has already been observed in previous projects, which indicated that the large dispersion of the data could be due to the inhomogeneity of the material. Regarding the irradiated state, the laboratories produced

T_0 values of 32 °C and 40 °C with mini-CT specimens, which were in good agreement with the resulting T_0 of 44 °C obtained using PCCv specimens.

Sugihara et al. [23] evaluated Japanese steel SFVQ1A subjected to a neutron fluence of 7.2×10^{19} n/cm² with mini-CT specimens and with 0.5T-CT specimens. Both data sets provided a valid T_0 of −1 °C for the mini-CT specimens and 8 °C for 0.5T-CT specimens. Thus, a difference of 9 °C was found, although the standard deviations (7.8 °C for mini-CT and 6 °C for 0.5T-CT) well overlapped each other.

After demonstrating the suitability of mini-CT specimens to characterize the JRQ material both in baseline and irradiated conditions [31], the same group (two laboratories) analyzed the possibility of evaluating the through-wall fracture toughness distribution with mini-CT specimens [32]. For this purpose, a block of JRQ steel was sliced into 13 layers, and the inner four layers (01J, 02J, 03J, and 04J) were used for the evaluation. The inner layer (01J) received a maximum neutron fluence of 5.38×10^{19} n/cm², which was attenuated to 2.54×10^{19} n/cm² at the 04J layer (60 mm from the inner surface). The irradiation temperature was 286 °C. Two more inner layers (01J, and 02J) were tested in the unirradiated conditions by one laboratory, providing (using mini-CTs) a T_0 of −121 °C and −115 °C for 01J and 02J, respectively. These values deviate by 3 °C and 8 °C, respectively, when compared to the results obtained with PCCv specimens. For the irradiated material, the T_0 results obtained by the two labs were comparable in layers 01J and 03J, but 02J showed a difference of 29 °C, which was close to $T_0 + 2\sigma$ (but still a bit larger) for the number of tested specimens. In conclusion, the use of mini-CT specimens was shown to be suitable to analyze the toughness distribution through the vessel wall. The use of specific fracture toughness values obtained at the inner surface (where neutron fluence is higher) may improve structural integrity assessments (e.g., pressurized thermal shock evaluations).

Ha et al. [35] also analyzed highly neutron-irradiated materials. In this case, three types of Japanese RPV steels were used, designated as Steel B, 3B, and 5B. Different levels of neutron fluence were applied: 11.3×10^{19} n/cm² for Steel B, 5.4×10^{19} n/cm² for 3B, and two fluence levels for 5B, the lower fluence (5BL) was 5.6×10^{19} n/cm² and the higher fluence (5BH) was 10.4×10^{19} n/cm². All the obtained T_0 values were valid. Finally, the specimen type effect was studied with Steel B in irradiated conditions: the T_0 value obtained through mini-CTs was about −12 °C, while in the case of PCCv specimens T_0 was −25 °C, resulting in a difference of 13 °C.

A collaborative program [25–27,42] was performed to characterize weld WF-70 material in both baseline and irradiated conditions. This low upper shelf Linde 80 weld had been previously characterized within the Heavy Section Steel Irradiation (HSSI) program with different types of larger C(T) specimens [43]. This program was performed in the 1990s, so the T_0 values reported here were recalculated from the original K_{Jc} data using the procedures of the current version of the ASTM E1921 [3] standard, obtaining T_0 values of −60 °C and 29 °C in unirradiated and irradiated conditions, respectively. The WF-70 material was then evaluated with mini-CT specimens by one laboratory in the unirradiated condition [12], resulting in a valid T_0 of −53 °C. The same laboratory evaluated the material in irradiated conditions [25], providing T_0 values of 2 °C or 12 °C, depending on the censoring criterion used for specimens with excessive ductile crack growth ($K_{Jclimit}$ vs. $K_{Jc\Delta a}$, see [25] for further details). Moreover, two additional laboratories tested the weld material by using the standard mini-CT specimen in irradiated conditions [27]. One of the laboratories could not obtain a valid T_0 value but provided a tentative T_{0Q} (provisional reference temperature [3]) of 31.3 °C. The other one obtained a valid T_0 of 34.8 °C. The combination of both data sets yielded a valid T_0 of 31.5 °C. All these data, in addition to those developed by a third laboratory, were analyzed as part of an inter-laboratory study [26]. T_0 in the irradiated condition varied between 13.2 °C and 17.5 °C, depending, again, on the censoring criterion.

Lambrecht et al. [9] investigated a series of mini-CT specimens taken from an A508-type weld metal in unirradiated and irradiated states. The steel was irradiated to a neutron fluence of 5×10^{19} n/cm² at 290 °C. In this study, modified mini-CT specimens with 20%

side grooves were employed. Fully valid T_0 values were obtained for the two conditions: −80.9 °C and −32.8 °C for unirradiated and irradiated conditions, respectively.

Uytdenhouwen et al. [33] studied the A508 Cl.2 RPV steel in irradiated and unirradiated conditions by means of mini-CT specimens. In this case, modified specimens were used, resulting in a T_0 of −88.1 °C for unirradiated material, which is similar to the −87.4 °C obtained with the 0.5T-CT specimen and to the −95.2 °C obtained with PCCv specimens.

Chen et al. [34] evaluated the applicability of mini-CT specimens to characterize the DBTR of two reduced activation ferritic martensitic (RAFM) steels proposed for fusion blanket applications, the EUROFER97 batch-3, and the F82H-BA12 steels. They did not obtain valid T_0 results, but they did observe that the resulting T_{0Q} was 40 °C lower for EUROFER97 batch-3 and 15 °C lower for F82H-BA12 when testing was performed on 0.5T-CT specimens with slanted fatigue precracks. In contrast, T_{0Q} was only 11 °C lower for EUROFER97 batch-3 and 4 °C higher for F82H-BA12 when testing was performed on mini-CT specimens with slanted fatigue precracks. One implication from this observation is that mini-CT specimens may be less sensitive to test imperfections and yield more consistent T_{0Q} values.

Han et al. [44] also studied the F82H RAFM steel, showing a negligible effect on the T_0 value of 1 °C when compared with the 1T-CT specimen.

Sokolov [45] studied KS-01 weld material in unirradiated and irradiated conditions. The T_0 derived from testing mini-CT specimens in the unirradiated condition was −9 °C, compared to −26 °C reported for a combination of 1T-CT, 0.5T-CT, and PCCv specimens. The T_0 temperature derived from irradiated mini-CTs was 153 °C, compared to the 139 °C reported for a combination of 1T-CT and 0.5T-CT specimens.

In summary, Figure 5 shows a comparison between the T_0 values obtained with large conventional specimens and those obtained using mini-CT specimens. The majority of the values gathered from the literature, including unirradiated and irradiated materials, are located between the bands of ±15 °C, as shown in the graph. On average, considering all the data reviewed here, the difference between T_0-mini and T_0-conventional specimens is less than −1 °C. Thus, in general, the values of T_0 obtained with mini-CT specimens are in good agreement with those obtained with larger specimens, demonstrating the robustness of mini-CT MC characterization.

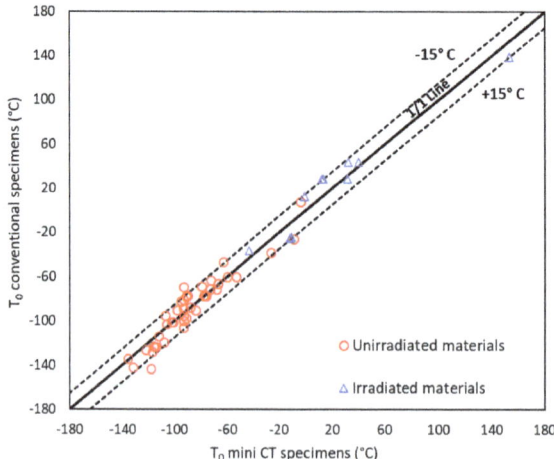

Figure 5. Comparison between T_0 values of unirradiated materials determined from conventional specimens and T_0 values obtained using mini-CT specimen. Results from: [7–9,11–15,18,20,22,23,25–27,29–31,33–35,37,38,40,42,44–46].

4. Regulatory Aspects

4.1. Initial MC Applications and Applicability Concerns

The size effect and scatter characterization of Wallin's MC concept were first introduced in 1984 [2,47], followed by Wallin's identification of a temperature dependence that works well for a wide range of ferritic steels in 1993 [48]. The first codification of MC techniques came with the 1997 adoption of the ASTM E1921 testing standard [3] and the 1999 ASME Code Cases that defined an MC-based reference temperature, RT_{T0}, that is the functional equivalent of RT_{NDT} [49–51]. RT_{T0} was moved from these code cases into the ASME Code itself in 2014. Early plant-specific applications of the MC in the USA included the Zion reactors in 1993 [52], and the Kewaunee reactor in 1998 [53,54]. MC was also used in South Korea to demonstrate the continued operating integrity of the Kori Unit 1 reactor in 2007 [55]. Continued regulatory use of MC techniques has occurred for PWR units in the USA having Linde 80 welds in their RPV beltline through the use of the "BAW-2308" approach [56], which allowed better estimation of the unirradiated RT_{NDT} value for Linde 80 welds using master curve results. The BAW-2308 approach has been used for nearly a dozen operating reactors.

During the late 1990s and early 2000s, there were no regulatory or Codes and Standards procedures for the use of MC. At that time, the regulator in the USA published a paper providing its views on MC technology and its applicability to the safety assessment of nuclear RPVs [57]. These authors stated that *"the Master Curve approach is promising . . . , however significant technical, process, and regulatory issues remain to be adequately addressed before full implementation of such an approach can be endorsed by the NRC."* The paper went on to highlight the following issues (direct quotations from [57] appear in italics):

1. *Fracture toughness characterization performed on the actual material in question or an appropriately qualified "surrogate".*
2. *Fracture toughness characterization performed in specimens with adequate constraint and at appropriate loading rates.*
3. *Quantification of the effects of irradiation on the shape (meaning temperature dependence) of the Master Curve.*
4. *Development and finalization of consensus Codes and Standards (ASTM, PVRC, and ASME.)*
5. *Revisions to USNRC rules and regulations governing RPV integrity.*

Item (1) was not directly related to the MC but rather to the long-standing recognition that some early construction plants in the USA did not monitor their limiting beltline material as part of their surveillance programs. The limiting materials were not monitored in some cases due to a lack of specificity in the then-current ASTM surveillance standard, which was attributable to a lack of knowledge in the 1960s and 1970s needed to identify the steels in the RPV beltline most sensitive to irradiation embrittlement. While item (1) also affected conventional RT_{NDT}-based assessments, it was raised by the NRC in the context of potential MC use due to the perception that using the MC would reduce some of the large implicit conservatisms thought to be inherent to RT_{NDT}. Item (1) has been largely resolved over the last 20 years, and moreover, systematic methods now exist to identify groups of similar materials from the large quantities of surveillance data now available [58].

Item (2) reflected concerns about (a) the use in T_0 estimation of what were then called "invalid" data (now referred to in ASTM E1921 and JEAC4216 as "censored" data) and (b) the change from using a dynamic Charpy test approach to a statically loaded fracture toughness test approach as the basis for positioning reference fracture toughness curves. Item (2a) has been addressed thoroughly via the standardization processes associated with ASTM E1921 and JEAC4216; consensus procedures are now well established to ensure the appropriate treatment of censored data. Item (2b) was addressed when the ASME Code moved in the late 1990s from using dynamic crack initiation and crack arrest data to static crack initiation data as the basis for the Code reference fracture toughness curve [59]. Moreover, there was the realization that the magnitude of Charpy shift with irradiation damage is linearly related to T_0 shift, and that the two quantities are generally equal, or nearly so [60]. To the extent that this question remains open, an ongoing effort to develop an

ASME Code Case aims to provide explicit procedures to address embrittlement, including treatment of uncertainties [60].

Item (3) has been resolved through extensive data analyses that show that the fixed Master Curve shape provides a good representation of the temperature dependence of RPV materials through any now foreseeable RPV lifetime [53,61], including steels with irradiation-induced shifts as high as $\Delta T_0 = 165\ °C$ [62]. While Wallin has presented a version of the MC having a temperature dependence related to both yield strength and T_0 [63], this variation is slight. Within the range of interest for RPV steels, the fixed shape MC continues to provide a good representation of the large collections of data now available. Neither MC testing standard, ASTM E1921 [3] or JEAC4216 [4], has seen the need to adopt the variable temperature dependence of MC from [63].

Concerning Item (4), as mentioned previously in ASTM E1921 [3] and JEAC4216 [4], MC testing standards are now well developed, representing over two decades of continuous improvement. ASME, JEAC, and other MC application standards continue to be developed, as summarized in Section 4.2.

Concerning Item (5), in the USA, the MC has not been adopted as part of any regulatory code or regulatory guidance document; however, the NRC has reviewed and endorsed parts of the ASME Code that use the MC as part of their ongoing processes that adopt the code and code cases as legal parts of the regulatory framework in the USA. The MC has seen legal endorsement by regulators in both Germany [64] and Switzerland [65]. In Japan, JEAC4206 has provisions for using the MC for the evaluation of plant operating limits [66]. While the 2016 version of JEAC4206 has been adopted by the Japan Electric Association, it has not yet been endorsed by the regulatory authority in Japan. The Swiss, German, and Japanese approaches are broadly similar to the ASME RT_{T0} approach, albeit differing in certain specific provisions and in their approaches to setting margins. Section 8.3 of [60] provides a more complete summary of the various approaches.

4.2. Current Activities in the USA

In the USA, considerable effort has been made since 2014 to incorporate MC concepts formally and comprehensively into the ASME Code, including a treatment of embrittlement and uncertainties, as follows:

- **Code Case N-830-1**: Following a 7.5-year development effort, Revision 1 to Code Case N830 was adopted into the Code in September 2021 following a unanimous and affirmative vote within ASME Section XI, including by the NRC representative [67]. This Code Case provides 5th percentile curves based on MC and extended MC models that can be used as alternatives to allowable toughness models now in the Code. Specifically, N-830 allows users to calculate, based only on knowledge of a T_0 value and the product form in question, the 5th percentile curves for 1T-K_{Jc} (which per this code case may be used as an alternate to K_{Ic}), K_{Ia}, J_{Ic}, and J-R. These toughness curves may be used in assessments of found flaws, to establish safe plant operating limits, and an evaluation of the fracture toughness needed on the upper shelf. The review and approval process associated with this code case included extensive interchange with the NRC to address the regulators' concerns on many topics, including validation and uncertainty treatment. This interchange is fully documented as an appendix to [67].
- **Code Case N-914**: This code case, which has been under development since 2019 and remains in draft form, provides a consistent and comprehensive methodology to assess embrittlement, including uncertainty treatment, for both conventional code approaches based on Charpy and NDT as well as MC-based approaches [60]. The current technical basis was reviewed by the NRC in 2021, and most questions were addressed. Future revisions of [60] will fully document the interchange with the NRC in the same manner as done with CC-N-830-1. In combination with Code Case N-830-1, Code Case N-914 provides a comprehensive and explicit methodology to use the MC and T_0 in ASME Code assessments.

In the reviews conducted on these two code cases, the NRC asked no questions concerning the use of mini-CT to estimate T_0. Perhaps the reason for this silence is that issues concerning the reliability of mini-CT-based T_0 values should be addressed through the ASTM balloting process for E1921, and that both CC-N-830-1 and CC-N-914 require the use of ASTM-valid T_0 values. A recent submittal to the NRC by the pressurized water reactor (PWR) Owners Group describes a planned project to collect an extensive database of T_0 values for PWR plants operating in the USA [58]. If executed, this project will include the testing of a substantial number of mini-CTs, the specimen being selected due to its frugal use of the limited archival and already irradiated materials that remain available. In their 15-page document requesting additional information [68] resulting from their review of [69], the NRC asked only one question about mini-CTs: *"discuss if additional uncertainty for mini-C(T) specimen data (i.e., uncertainly greater than what would be applied for larger C(T) specimens) would be included in the adjustment or margin terms."*

In summary, the application of mini-CT data in regulatory analysis remains a relatively new development, and this may explain the present lack of detailed questions from regulators about the mini-CT that extend beyond the much larger body of questions that regulators have asked about the MC over the past two to three decades.

5. Conclusions

The use of mini-CT specimens for master curve testing is a good option to reuse previously tested samples. It provides options for plants lacking sufficient amounts of material to continue to use conventional surveillance programs, during long-term operation. In particular, there is no bias in the reference temperature T_0 determined by testing mini-CTs when compared to larger specimens (see Figure 5). Nevertheless, certain aspects associated with mini-CT testing present unique challenges. In particular, the selection of appropriate test temperatures is crucial because the reduced measuring capacity of mini-CT specimens ($K_{Jclimit}$) limits the test temperature range over which censoring is unlikely. As a consequence, a higher number of tests (N≈12) is often required for the determination of a valid T_0 value using mini-CTs. This can be addressed in future revisions of the related standards [3,4]. In addition, even a modification of the 50 °C exclusion criterion could be considered by not invalidating data at very low test temperatures (i.e., $T < T_0 - 50$ °C).

Additional data and/or analyses will help to underpin the findings already discussed in this paper and lead to their resolution. Specific topics currently under consideration include the following:

- Censoring statistics (i.e., censoring probability as a function of test temperature)
- Re-evaluation of the censoring criterion for low upper shelf materials (cf. Equation (4)).
- Effects of side grooving and specific benefits.
- Effect of inhomogeneities and how they are handled in standards.

Author Contributions: Conceptualization, M.S. and S.C.; methodology, M.S. and S.C.; formal analysis, M.S., S.C., M.K., E.A., W.S. and M.Y.; investigation, M.S., S.C., M.K., E.A., W.S. and M.Y.; writing—original draft preparation, M.S., S.C., M.K., E.A., W.S. and M.Y.; writing—review and editing, M.S., S.C., M.K., E.A., W.S. and M.Y. All authors have read and agreed to the published version of the manuscript.

Funding: This project received funding from the Euratom Research & Training Programme 2019–2020 under grant agreement No. 900014 (FRACTESUS).

Data Availability Statement: Not applicable.

Conflicts of Interest: The authors declare no conflict of interest.

References

1. Wallin, K. Master curve analysis of the "Euro" fracture toughness dataset. *Eng. Fract. Mech.* **2002**, *69*, 451–481. [CrossRef]
2. Wallin, K. The scatter in KIC-results. *Eng. Fract. Mech.* **1984**, *19*, 1085–1093. [CrossRef]
3. ASTM E1921; Standard Test Method for Determination of Reference Temperature, T0, for Ferritic Steels in the Transition Range. ASTM International: West Conshohocken, PA, USA, 2021. [CrossRef]
4. JEAC4216-2015; Test Method for Determination of Reference Temperature T0 of Ferritic Steels. Japan Electric Association Code. Electric Association: Tokyo, Japan, 2015.
5. Brynk, T.; Uytdenhouwen, I.; Obermeier, F.; Altstadt, E.; Kopriva, R.; Serrano, M.; Arffman, P. Fractesus project: Final selection of RPV materials for unirradiated and irradiated round robins. *Am. Soc. Mech. Eng. Press. Vessel. Pip. Div. PVP* **2022**, *1*, 1–11. [CrossRef]
6. Cicero, S.; Lambrecht, M.; Swan, H.; Arffman, P.; Altstadt, E.; Petit, T.; Obermeier, F.; Arroyo, B.; Álvarez, J.A.; Lacalle, R. Fracture mechanics testing of irradiated RPV steels by means of sub-sized specimens: FRACTESUS project. *Procedia Struct. Integr.* **2020**, *28*, 61–66. [CrossRef]
7. Miura, N.; Soneda, N. Evaluation of Fracture Toughness by Master Curve Approach Using Miniature C(T) Specimens. In Proceedings of the ASME 2010 Pressure Vessels and Piping Division/K-PVP Conference, Bellevue, WA, USA, 18–22 July 2010; ASME: New York, NY, USA; Volume 1, pp. 593–602. [CrossRef]
8. Chaouadi, R.; Van Walle, E.; Scibetta, M.; Gérard, R. On the use of miniaturized ct specimens for fracture toughness characterization of RPV materials. *Am. Soc. Mech. Eng. Press Vessel Pip. Div. PVP* **2016**, *1B*, 1–10. [CrossRef]
9. Lambrecht, M.; Chaouadi, R.; Uytdenhouwen, I.; Gérard, R. Fracture toughness characterization in the transition and ductile regime of an a508 type weld metal with the mini-ct geometry before and after irradiation. *Am. Soc. Mech. Eng. Press. Vessel. Pip. Div. PVP* **2020**, *1*, 1–7. [CrossRef]
10. ASTM E399; Standard Test Method for Linear-Elastic Plane-Strain Fracture Toughness of Metallic Material. ASTM International: West Conshohocken, PA, USA, 2022; vol. 03.01. [CrossRef]
11. Scibetta, M.; Lucon, E.; Van Walle, E. Optimum use of broken Charpy specimens from surveillance programs for the application of the master curve approach. *Int. J. Fract.* **2002**, *116*, 231–244. [CrossRef]
12. Sokolov, M.A. Use of mini-CT specimens for fracture toughness characterization of low upper-shelf linde 80 weld. *Am. Soc. Mech. Eng. Press. Vessel. Pip. Div. PVP* **2017**, *1A*, 20–23. [CrossRef]
13. Yamamoto, M.; Kimura, A.; Onizawa, K.; Yoshimoto, K.; Ogawa, T.; Chiba, A.; Hirano, T.; Sugihara, T.; Sugiyama, M.; Miura, N. A round robin program of Master Curve evaluation using miniature C(T) specimens: First round robin test on uniform specimens of reactor pressure vessel material. *Am. Soc. Mech. Eng. Press. Vessel. Pip. Div. PVP* **2012**, *6*, 73–79. [CrossRef]
14. Yamamoto, M.; Kimura, A.; Onizawa, K.; Yoshimoto, K.; Ogawa, T.; Mabuchi, Y.; Viehrig, H.W.; Miura, N.; Soneda, N. A round Robin program of master curve evaluation using miniature C(T) specimens-3RD report: Comparison of T0 under various selections of temperature conditions. *Am. Soc. Mech. Eng. Press. Vessel. Pip. Div. PVP* **2014**, *1*, 1–7. [CrossRef]
15. Yamamoto, M.; Onizawa, K.; Yoshimoto, K.; Ogawa, T.; Mabuchi, Y.; Miura, N. Round Robin Program of Master Curve Evaluation Using Miniature C(T) Specimens—2nd Report: Fracture Toughness Comparison in Specified Loading Rate Condition. *Am. Soc. Mech. Eng. Press. Vessel. Pip. Div. PVP* **2013**, *1*, 1–8. [CrossRef]
16. Landes, J.D. J calculation from front face displacement measurement on a compact specimen. *Int. J. Fract.* **1980**, *16*, R183–R185. [CrossRef]
17. Miura, N.; Momoi, Y.; Yamamoto, M. Relation Between Front-Face and Load-Line Displacements on a C(T) Specimen by Elastic-Plastic Analysis. *Am. Soc. Mech. Eng. Press. Vessel. Pip. Div. PVP* **2015**, 1–6. [CrossRef]
18. Yamamoto, M.; Miura, N. Applicability of miniature C(T) specimens for the Master Curve evaluation of RPV weld metal. *Am. Soc. Mech. Eng. Press. Vessel. Pip. Div. PVP* **2015**, *1A*, 1–7. [CrossRef]
19. Wallin, K.; Yamamoto, M.; Ehrnstén, U. Location of Initiation Sites in Fracture Toughness Testing Specimens: The Effect of Size and Side Grooves. *Am. Soc. Mech. Eng. Press. Vessel. Pip. Div. PVP* **2016**, *1*, 1–9. [CrossRef]
20. Lambrecht, M.; Chaouadi, R.; Li, M.; Uytdenhouwen, I.; Scibetta, M. On the possible relaxation of the ASTM E1921 and ASTM E1820 Standard specifications with respect to the use of the mini-CT specimen. *Mater. Perform. Charact.* **2020**, *9*, 593–607. [CrossRef]
21. Lindqvist, S.; Kuutti, J. Sensitivity of the Master Curve reference temperature T0 to the crack front curvature. *Theor. Appl. Fract. Mech.* **2022**, *122*, 103558. [CrossRef]
22. Tobita, T.; Nishiyama, Y.; Ohtsu, T.; Udagawa, M.; Katsuyama, J.; Onizawa, K. Fracture Toughness Evaluation of Reactor Pressure Vessel Steels by Master Curve Method Using Miniature Compact Tension Specimens. *J. Press. Vessel Technol.* **2015**, *137*, 4–11. [CrossRef]
23. Sugihara, T.; Hirota, T.; Sakamoto, H.; Yoshimoto, K.; Tsutsumi, K.; Murakami, T. Applicability of miniature C(T) specimen to fracture toughness evaluation for the irradiated japanese reactor pressure vessel steel. *Am. Soc. Mech. Eng. Press. Vessel. Pip. Div. PVP* **2017**, *1*, 1–8. [CrossRef]
24. Sokolov, M.A.; Nanstad, R.K. Comparison of irradiation-induced shifts of KJc and Charpy impact toughness for reactor pressure vessel steels. *ASTM Spec. Tech. Publ.* **1999**, *1325*, 167–190.
25. Sokolov, M.A. Use of mini-CT specimens for fracture toughness characterization of low upper-shelf linde 80 weld before and after irradiation1. *Am. Soc. Mech. Eng. Press. Vessel. Pip. Div. PVP* **2018**, *1*, 1–6. [CrossRef]

26. Server, W.; Sokolov, M.; Yamamoto, M.; Carter, R. Inter-Laboratory Results and Analyses of Mini-C(T) Specimen Testing of an Irradiated Linde 80 Weld Metal. *Am. Soc. Mech. Eng. Press. Vessel. Pip. Div. PVP* **2018**, *1*, 1–5. [CrossRef]
27. Yamamoto, M. Trial study of the master curve fracture toughness evaluation by mini-C(T) specimens for low upper shelf weld metal linde-80. *Am. Soc. Mech. Eng. Press. Vessel. Pip. Div. PVP* **2018**, *1*, 1–8. [CrossRef]
28. Yamamoto, M.; Kirk, M.; Shinko, T. Master curve evaluation using the fracture toughness data at low test temperature of T-T0<-50 °C. *Am. Soc. Mech. Eng. Press. Vessel. Pip. Div. PVP* **2022**, *1*, 1–8. [CrossRef]
29. Yamamoto, M.; Miura, N. Applicability of miniature-c(T) specimen for the master curve evaluation of rpv weld metal and heat affected zone. *Am. Soc. Mech. Eng. Press. Vessel. Pip. Div. PVP* **2016**, *1*, 1–8. [CrossRef]
30. Ha, Y.; Tobita, T.; Takamizawa, H.; Nishiyama, Y. Fracture toughness evaluation of neutron-irradiated reactor pressure vessel steel using miniature-C(T) specimens. *Am. Soc. Mech. Eng. Press. Vessel. Pip. Div. PVP* **2017**, *1*, 1–5. [CrossRef]
31. Yamamoto, M. The Master Curve Fracture Toughness Evaluation of Irradiated Plate Material JRQ Using Miniature-C(T) Specimens. *Am. Soc. Mech. Eng. Press. Vessel. Pip. Div. PVP.* **2017**, *1*, 1–8. [CrossRef]
32. Yamamoto, M.; Kobayashi, T. Evaluation of Through Wall Fracture Toughness Distribution of IAEA Reference Material JRQ by Mini-C(T) Specimens and the Master Curve Method. *Am. Soc. Mech. Eng. Press. Vessel. Pip. Div. PVP.* **2018**, *1*, 1–8. [CrossRef]
33. Uytdenhouwen, I.; Chaouadi, R. Effect of neutron irradiation on the mechanical properties of an a508 cl.2 forging irradiated in a bami capsule. *Am. Soc. Mech. Eng. Press. Vessel. Pip. Div. PVP.* **2020**, *1*, 2–10. [CrossRef]
34. Chen, X.F.; Sokolov, M.A.; Gonzalez De Vicente, S.M.; Katoh, Y. Specimen Size and Geometry Effects on the Master Curve Fracture Toughness Measurements of EUROFER97 and F82H Steels. *Am. Soc. Mech. Eng. Press. Vessel. Pip. Div. PVP.* **2022**, *1*, 1–9. [CrossRef]
35. Ha, Y.; Tobita, T.; Ohtsu, T.; Takamizawa, H.; Nishiyama, Y. Applicability of Miniature Compact Tension Specimens for Fracture Toughness Evaluation of Highly Neutron Irradiated Reactor Pressure Vessel Steels. *J. Press. Vessel Technol.* **2018**, *140*, 1–6. [CrossRef]
36. Yamamoto, M.; Sakuraya, S.; Kitsunai, Y.; Kirk, M. Practical procedure of test temperature selection for mini-C(T) master curve evaluation. *Am. Soc. Mech. Eng. Press. Vessel. Pip. Div. PVP* **2022**, *1*, 1–7. [CrossRef]
37. Takamizawa, H.; Tobita, T.; Ohtsu, T.; Katsuyama, J.; Nishiyama, Y.; Onizawa, K. Finite element analysis on the application of MINI-C(T) test specimens for fracture toughness evaluation. *Am. Soc. Mech. Eng. Press. Vessel. Pip. Div. PVP* **2015**, *1A*, 1–7. [CrossRef]
38. Sánchez, M.; Cicero, S.; Arroyo, B.; Cimentada, A. On the Use of Mini-CT Specimens to Define the Master Curve of Unirradiated Reactor Pressure Vessel Steels with Relatively High Reference Temperatures. *Theoretical and Applied Fracture Mechanics* **2023**, *124*, 103736. [CrossRef]
39. Hall, J.B.; Yoon, K.K. Quasi-Static Loading Rate Effect on the Master Curve Reference Temperature of Ferritic Steels and Implications. *Am. Soc. Mech. Eng. Press. Vessel. Pip. Div. PVP* **2003**, *1*, 9–14. [CrossRef]
40. Yamamoto, M.; Carter, R.; Viehrig, H.; Lambrecht, M. A Round Robin Program of Master Curve Evaluation using Miniature C (T) Specimens (Comparison of T0 for a Weld Metal). 20–25 August 2017.
41. International Atomic Energy Agency (IAEA). *Application of Surveillance Programme Results to Reactor Pressure Vessel Integrity Assessment. Results of a Coordinated Research Project 2000–2004*; International Atomic Energy Agency (IAEA): Vienna, Austria, 2005; Volume 1556.
42. Ickes, M.R.; Brian Hall, J.; Carter, R.G. Fracture toughness characterization of low upper-shelf linde 80 weld using mini-C(T) specimens. *Am. Soc. Mech. Eng. Press. Vessel. Pip. Div. PVP* **2018**, *1*, 1–6. [CrossRef]
43. McCabe, D.; Nanstad, R.; Iskander, S.; Heatherly, D.; Swain, R. NUREG/CR-5736; *Evaluation of WF-70 Weld Metal from the Midland Unit 1 Reactor Vessel*; U.S. Nuclear Regulatory Commission: Rockville, MD, USA, 2000.
44. Han, W.; Yabuuchi, K.; Kasada, R.; Kimura, A.; Wakai, E.; Tanigawa, H.; Liu, P.; Yi, X.; Wan, F. Application of small specimen test technique to evaluate fracture toughness of reduced activation ferritic/martensitic steel. *Fusion Eng. Des.* **2017**, *125*, 326–329. [CrossRef]
45. Sokolov, M.A. Use of mini-CT specimens for fracture toughness characterization of irradiated highly embrittled weld. *Am. Soc. Mech. Eng. Press. Vessel. Pip. Div. PVP* **2022**, *1*, 10–13. [CrossRef]
46. Zhou, Z.; Tong, Z.; Qian, G.; Berto, F. Specimen Size Effect on the Ductile-Brittle Transition Reference Temperature of A508-3 Steel. *Theoretical and Applied Fracture Mechanics* **2019**, *104*, 102370. [CrossRef]
47. Wallin, K.; Saario, T.; Törrönen, K. Statistical Model for Carbide Induced Brittle Fracture in Steel. *Met. Sci.* **1984**, *18*, 13–16. [CrossRef]
48. Wallin, K. Irradiation Damage Effects on the Fracture Toughness Transition Curve Shape for Reactor Vessel Steels. *Int. J. Pres. Vessel. Pip.* **1993**, *55*, 61–79. [CrossRef]
49. ASME Boiler and Pressure Vessel Code Case N-629. *Use of Fracture Toughness Test Data to Establish Reference Temperature for Pressure Retaining Materials*; Section XI, Division 1; ASME: New York, NY, USA, 1999.
50. ASME Boiler and Pressure Vessel Code Case N-631. *Use of Fracture Toughness Test Data to Establish Reference Temperature for Pressure Retaining Materials Other Than Bolting for Class 1 Vessels Section III, Division 1*; ASME: New York, NY, USA, 1999.
51. Electric Power Research Institute. *Application of Master Curve Fracture Toughness Methodology for Ferritic Steels (PWRMRP-01): PWR Materials Reliability Project (PWRMRP)*; EPRI: Palo Alto, CA, USA, 1999; TR-108390 Revision 1.

52. Yoon, K.K. *Fracture Toughness Characterization of WF-70 Weld Metal*; Report to the B&W Owners Group Materials Committee, BAW-2202; Babcock and Wilcox Company: Nuclear Power Division, Virginia, September 1993.
53. Lott, R.G.; Kirk, M.T.; Kim, C.C. Master Curve Strategies for RPV Assessment. Westinghouse Electric Company: Pittsburgh, PA, USA, WCAP-15075. September 1998; Available on the USNRC website at Legacy ADAMS 9811240260 and at ADAMS ML111861647.
54. NRC Safety Evaluation Report on Kewaunee Master Curve Submittal, Letter of 1st May 2001 from Lamb to Reddemann, ADAMS ML011210180.
55. Lee, B.S.; Hong, J.H.; Lee, D.H.; Choi, D.G. RTPTS Re–Evaluation of Kori–1 Rpv Beltline Weld By Master Curve Tests. In Proceedings of the Second International Symposium on Nuclear Power Plant Life Management, Shanghai, China, 15–18 October 2007. Paper Number IAEA-CN-155-063.
56. Framatome ANP, Inc. Initial RT_{NDT} of Linde-80 Weld Materials. Report to the PWR Owners Group, BAW-2308 Rev. 2A. March 2008.
57. Mayfield, M.; Vassilaros, M.; Hackett, E.; Wichman, K.; Strosnider, J.; Shao, L. Application of revised fracture toughness curves in pressure vessel integrity analysis. In Proceedings of the Transactions of the 14th International Conference on Structural Mechanics in Reactor Technology (SMiRT 14), Lyon, France, 12–22 August 1997. G01/2.
58. Kirk, M.; Hashimoto, Y.; Nomoto, A. Application of a Machine Learning Approach Based on Nearest Neighbors to Extract Embrittlement Trends from RPV Surveillance Data. *J. Nucl. Mater.* **2022**, *568*, 153886. [CrossRef]
59. Bamford, W.; Stevens, G.; Griesbach, T.; Malik, S. *The 2000 Technical Basis for Revised P-T Limit Curve Methodology*; ASME Pressure Vessel and Piping Meeting: Seattle, WA, USA, 2000.
60. Electric Power Research Institute. *Methods to Address the Effects of Irradiation Embrittlement in Section XI of the ASME Code: Estimation of an Irradiated Reference Temperature Using either Traditional Charpy Approaches or Master Curve Data*; EPRI: Palo Alto, CA, USA, 2021; p. 3002020911.
61. Kirk, M. *The Technical Basis for Application of the Master Curve to the Assessment of Nuclear Reactor Pressure Vessel Integrity*; ADAMS ML093540004; USA Nuclear Regulatory Commission: Washington, DC, USA, 2009.
62. Sokolov, M.; Nanstad, R.; Remec, I.; Baldwin, C.; Swain, R. *Fracture Toughness of an Irradiated, Highly Embrittled Reactor Pressure Vessel Weld*; Oak Ridge National Laboratory: Oak Ridge, TN, USA, 2003; ORNL/TM-2002/293.
63. Wallin, K. The Elusive Temperature Dependence of the Master Curve. In Proceedings of the 13th International Conference on Fracture 2013, ICF-13, Beijing, China, 16–21 June 2013; 2013; Volume 7, pp. 5609–5617.
64. *KTA-3203*; Surveillance of the Irradiation Behavior of Reactor Pressure Vessel Materials of LWR Facilities. Safety Standards of the Nuclear Safety Standards Commission (KTA): Salzgitter, Germany, 2011.
65. *ENSI-B01/d*; Alterungsüberwachung, Richtlinie für die Schweizerischen Kernanlagen. Swiss Federal Nuclear Safety Inspectorate ENSI: Brugg, Switzerland, 2012.
66. *JEAC 4206-2016*; Method of Verification Tests of the Fracture Toughness for Nuclear Power Plant Components. Japan Electric Association Code. Electric Association: Tokyo, Japan, 2016.
67. Electric Power Research Institute. *Technical Basis for ASME Code Case N-830-1, Revision 1 (MRP-418, Revision 1): Direct Use of Master Curve Fracture Toughness Curve for Pressure-Retaining Materials of Class 1 Vessels, Section XI*; EPRI: Palo Alto, CA, USA, 2019; p. 3002016008.
68. U.S. Nuclear Regulatory Commission's request for additional information by the office of nuclear reactor regulation on topical report PWROG-18068-NP, revision 1. Use of Direct Fracture Toughness for Evaluation of RPV Integrity for the Pressurized Water Reactor Owners Group Project No. 99902037; EPID: L-2021-TOP-0027, ADAMS ML2208A246. USA Nuclear Regulatory Commission: Washington, DC, USA, 2021.
69. Hall, J.B. *Use of Direct Fracture Toughness for Evaluation of RPV Integrity*; PWROG-18068-Np-Rev. 1; Westinghouse Electric Company: Pittsburgh, PA, USA, July 2021.

Disclaimer/Publisher's Note: The statements, opinions and data contained in all publications are solely those of the individual author(s) and contributor(s) and not of MDPI and/or the editor(s). MDPI and/or the editor(s) disclaim responsibility for any injury to people or property resulting from any ideas, methods, instructions or products referred to in the content.

Review

Twin-Related Grain Boundary Engineering and Its Influence on Mechanical Properties of Face-Centered Cubic Metals: A Review

Xiaowu Li [1,2,*], Xianjun Guan [1,3], Zipeng Jia [1,3], Peng Chen [1,3], Chengxue Fan [1,3] and Feng Shi [1,3]

1. Department of Materials Physics and Chemistry, School of Materials Science and Engineering, Northeastern University, Shenyang 110819, China
2. State Key Laboratory of Rolling and Automation, Northeastern University, Shenyang 110819, China
3. Key Laboratory for Anisotropy and Texture of Materials, Ministry of Education, Northeastern University, Shenyang 110819, China
* Correspondence: xwli@mail.neu.edu.cn

Abstract: On the basis of reiterating the concept of grain boundary engineering (GBE), the recent progress in the theoretical models and mechanisms of twin-related GBE optimization and its effect on the mechanical properties is systematically summarized in this review. First, several important GBE-quantifying parameters are introduced, e.g., the fraction of special grain boundaries (GBs), the distribution of triple-junctions, and the ratio of twin-related domain size to grain size. Subsequently, some theoretical models for the GBE optimization in face-centered cubic (FCC) metals are sketched, with a focus on the model of "twin cluster growth" by summarizing the in-situ and quasi-in-situ observations on the evolution of grain boundary character distribution during the thermal-mechanical process. Finally, some case studies are presented on the applications of twin-related GBE in improving the various mechanical properties of FCC metals, involving room-temperature tensile ductility, high-temperature strength-ductility match, creep resistance, and fatigue properties. It has been well recognized that the mechanical properties of FCC materials could be obviously improved by a GBE treatment, especially at high temperatures or under high cyclic loads; under these circumstances, the materials are prone to intergranular cracking. In short, GBE has tremendous potential for improving the mechanical properties of FCC metallic materials, and it is a feasible method for designing high-performance metallic materials.

Keywords: grain boundary engineering; mechanical property; face-centered cubic metal; special grain boundary; annealing twin

1. Introduction

Early in the 1880s, Sorby first observed with an optical microscope that the microstructure of a blister steel was composed of numerous grains of various shapes and the grain boundaries (GBs) between adjoining grains. Since then, materials researchers have paid an increasing amount of attention to the GBs and interfaces (including phase boundaries) to explore a well-established method in materials design and performance improvement [1]. After the past ~140 years of study, the understanding of GBs and interfaces has significantly improved. It is now well recognized that the GB is an important component of the microstructure in polycrystalline materials and that the number, type, and distribution of GBs play critical roles in the materials' properties [2–6].

In addition, when it comes to the mechanical properties, GBs can act as the main obstacle to dislocation slip during plastic deformation, and thus become important sources of strength and work hardening of polycrystalline metallic materials [7,8]. Meanwhile, GBs may also be the preferred location for crack nucleation due to the weakened bonding strength between the atoms on both sides of the structurally disordered interface and the higher stress concentration derived from the pile-up of dislocations [9–13]. In addition, most noteworthy, the structural order of various GBs is significantly different [14–16], so that the capacity of various GBs to resist intergranular cracks is also different [10,17–20]. Therefore, cracking is most likely to occur during plastic deformation at or along ordinary random high-angle GBs (RHAGBs) with higher structural disorder and interface energy, while special GBs with low energy (will be introduced in detail later) can generally maintain a high resistance to cracking [10,11,13,21,22].

Therefore, many researchers have made great efforts to reveal the influence of grain boundary character distribution (GBCD) optimization (also known as grain boundary engineering, or GBE) on the mechanical properties, e.g., tensile property, creep resistance, and fatigue resistance [6,12,13,23–25], and some praiseworthy research findings have been achieved in this context. While the study on the GBE approach to improving the mechanical properties of metallic materials is in progress, on the basis of further clarifying the concept of GBE, this review focuses on the latest progress in the theoretical models and mechanisms of GBE optimization and its impact on the mechanical properties. It is hoped that the summary of the recent studies of GBE may provide some valuable references for the development of advanced metallic materials that exhibit high resistances to GBE-related damage in their practical applications.

2. Twin-Related GBE

The main idea of GBE evolves from the concept of "GB design and control" proposed by Watanabe [2,3,26]. In addition, after nearly 50 years of development, GBE has emerged as a mature method that can be applied to a variety of metallic materials, such as copper alloys [6,12,27], nickel-based alloys [28–31], austenitic stainless steels [9,10,20,32,33], and lead alloys [34,35]. These materials generally have the common characteristic that their stacking fault energies are relatively low and annealing twins (ATs) are easily formed during the thermal-mechanical process (TMP); based on this, the so-called method of twin-related GBE has been well developed.

The twin-related GBE is to induce a large number of AT boundaries (or ATs), namely $\Sigma 3$ GBs, in face-centered cubic (FCC) metals by the means of thermal-mechanical treatment, and other low-Σ coincidence site lattice (CSL) GBs can be further induced through the mutual interactions between two annealing twins or even between ATs and ordinary RHAGBs, thus blocking the connectivity of RHAGBs [6,36–39]. Additionally, some previous studies [10,20,40,41] have revealed that the low-ΣCSL GBs introduced by twin-related GBE exhibit a high degree of structural stability and are not prone to second-phase precipitation during medium- to high temperature annealing or welding processes. Thus, the low-ΣCSL GBs, especially in the welding heat affected zone, often exhibit a higher corrosion resistance in corrosive environments compared with ordinary RHAGBs [42]. Moreover, many investigations [9,10,17,21,43] have substantiated that twin-related GBE is an effective way to improve the intergranular corrosion resistance, intergranular stress corrosion cracking resistance, and other properties that are closely related to GBs in some FCC metals.

Consequently, the key to the realization of GBE is to induce the formation of as many annealing twins as possible during the TMP. In addition, it should be noted that only in FCC metallic materials deformed by planar-slipping of dislocations can a large number of ATs be easily induced [6]. Therefore, the twin-related GBE can only be availably applied to some FCC metallic materials with low stacking fault energy or numerous short-range order structures [6,9,41,44–46]. Many important engineering materials, such as austenitic stainless steel, nickel-based alloys, and copper alloys, are involved in this kind of material, so the twin-related GBE has received widespread attention.

3. GBE-Quantifying Parameters

As mentioned above, the main purpose of twin-related GBE is to optimize the GBCD in materials, namely, increasing the fraction of low-ΣCSL GBs and blocking the connectivity of RHAGBs. In this case, it is significantly important to know how to quantify the degree of GBCD optimization. To this end, several important GBE-quantifying parameters have been successively proposed, as summarized below.

3.1. Fraction of Special GBs

The fraction of special GBs (f_{SBs}) is defined by the ratio of the length of special GBs to the total length of all GBs [20], which was most widely used in the evaluation of GBE optimization. In the statistics of f_{SBs}, the CSL GBs with $\Sigma \leq 29$ are generally regarded as the special GBs [47–49], since these types of GBs have lower energy [7,50,51], and exhibit some unique performances, e.g., lower diffusivity, lower resistivity, lower sensitivity to solute atom segregation, and higher resistance to GB sliding and crack initiation [7].

Furthermore, with the deepening of GBE studies, it has been recognized that, apart from the f_{SBs}, the role of special GBs is also closely related to their location, i.e., out of or in networks of random GBs. Lehockey et al. [52] reported that the coherent $\Sigma 3$ GBs were dominant in all special GBs in FCC metals with low stacking fault energy; however, most of these coherent $\Sigma 3$ GBs were localized outside of RHAGB networks and cannot make any positive contributions to the tailoring of RHAGB networks. Naturally, it is difficult for these coherent $\Sigma 3$ GBs to impede the intergranular stress corrosion along RHAGBs. Therefore, Lehockey et al. [52] proposed the concept of effective special GBs and suggested that only the special GBs that can block the network connectivity of the random high-angle GBs are effective in preventing the failure of materials along GBs. However, the difference in the anti-cracking properties of the same type of special GBs was even sometimes reported [20,53], so that there was as yet no unified opinion on the definition of effective special GBs. Further, it may be related to the fact that the interface index is still not considered in the current electron backscatter diffraction (EBSD) characterization. Thus, the further developments in the theory of GBs and the relevant characterization methods need to be emphasized for the better application of GBE.

3.2. Distribution of Triple-Junctions

The ultimate purpose of twin-related GBE treatment is to interrupt the network connectivity of RHAGBs and thus hinder crack propagation along RHAGBs. Therefore, how to quantify the blocking of RHAGB by the GBCD optimization is of great significance for GBE. Kumar et al. [54] proposed that the connectivity of RHAGBs can be evaluated by the statistics on the distribution of triple-junctions, which are classified as different types according to the number of special GBs they contain. For example, triple-junctions with zero, one, two, and three low-energy special CSL GBs are classified as J0, J1, J2, and J3 junctions, respectively. Among these triple-junctions, the RHAGBs are interconnected with each other at the J0 and J1 junctions, and thereby cracks can pass through the junctions without hindrance and propagate along the RHAGBs. In contrast, J3 junctions are generally too stable to meet the conditions for crack nucleation and propagation. Only at the J2 junctions can the cracks be captured by special GBs. Hence, the capture probability of cracks can be quantified by statistically calculating the distribution of $f_{J2}/(1-f_{J3})$, where f_{J2} and f_{J3} represent the proportion of J2 and J3 junctions, respectively. Several experimental studies [6,20,55] have confirmed that there indeed exists a strongly positive correlation between the $f_{J2}/(1-f_{J3})$ and the blocking degree of RHAGB connectivity in FCC metals.

3.3. Ratio of Twin Related Domain Size to Grain Size

In the study on the mechanism of twin-related GBE, it has recently been realized that the formation of annealing twins during TMP plays an important role in increasing the f_{SBs} and interrupting the network connectivity of RHAGBs, since efficiently inducing the formation of ATs can not only directly increase the fraction of special GBs but also be beneficial to decreasing the grain size. Furthermore, the quasi-in situ EBSD observation on the evolution of GBCD further indicated that the formation and growth of ATs indeed played a critical role in GBCD optimization [36]. For example, the f_{SBs} in a GBE- treated material are directly related to the number of ATs in a twin-related domain (TRD), which is defined as Σ3ⁿ twin cluster (namely, a large cluster of spatially adjacent twin-related grains). In light of this, Barr et al. [36] suggested that the ratio v of TRD size to grain size should be another important indicative indicator of GBE in polycrystalline materials.

4. Mechanism of GBCD Optimization—"Twin Cluster Growth" Model

At the turn of the century, the study on the mechanism of GBCD optimization in FCC metallic materials has drawn immense attention from researchers, and several theoretical models for GBE have been proposed in succession, as shown in Figure 1. As a result, the "Σ3 GB regeneration model" [56] indicates that the coherent AT boundaries in different recrystallized grains can interact with each other to induce the formation of Σ9 GBs, and then the mobile Σ9 GBs further interact with some other Σ3 GBs, thus inducing the formation of other low-energy SBs (Figure 1a). The "high ΣCSL GB decomposition model" [57] suggests that the low-ΣCSL SBs can be derived from the decomposition of high-ΣCSL GBs (Figure 1b). According to the "special fragment model" [17], the GBCD optimization is mainly realized by the SB fragments caused by AT emitting in the RHAGB network (Figure 1c). The "incoherent Σ3 GB migration model" [58] indicates that the formation of SBs is mainly achieved by the migration of incoherent Σ3 GBs (Figure 1d). However, even though these models can explain some laws of GBCD evolution in FCC metals with low stacking fault energy to a certain extent, there still exist some obvious inadequacies due to the lack of understanding of the microstructure evolution during the TMP. On the basis of summarizing the recent in situ or quasi-in situ observations on the microstructure evolution of FCC metals, the "twin cluster growth" model is further introduced below.

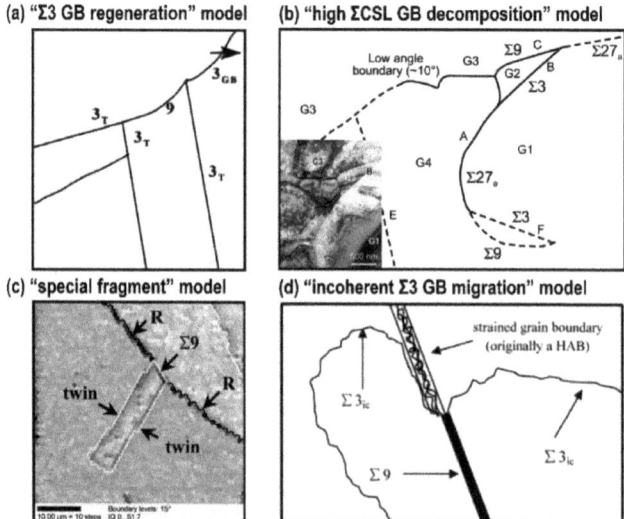

Figure 1. Theoretical models for the GBCD optimization in FCC metals Adopted from Refs. [17,56–58].

Additionally, through the quasi-in situ or in situ observation of the GBCD evolution during the TMP of some FCC metals (e.g., 304, 316L austenitic stainless steels, and copper alloys) [6,36,59], it has been well recognized that the microstructural evolution during GBE treatment is mainly completed by strain-induced GB migration. For example, the ordinary RHAGBs driven by the stored strain energy migrate from the twin clusters to the deformed matrix during the annealing process of the TMP, and ATs are constantly nucleated behind the migrating GBs and grow up with the migration of GBs, as shown in Figure 2. Therefore, inducing the nucleation of as many ATs as possible is crucial to realizing the optimization of GBCD in the process. First, the nucleation of ATs, as mentioned above, can directly increase the f_{SBs}. Second, ATs with different orientations nucleated behind RHAGBs may interact with each other, thus inducing other low-energy CSL special GBs (see Figure 2b,c). Finally, the formation of ATs can induce the structural transition of migrating RHAGBs from disorder to order, which are implanted in the network of RHAGBs and thus interrupt the network connectivity.

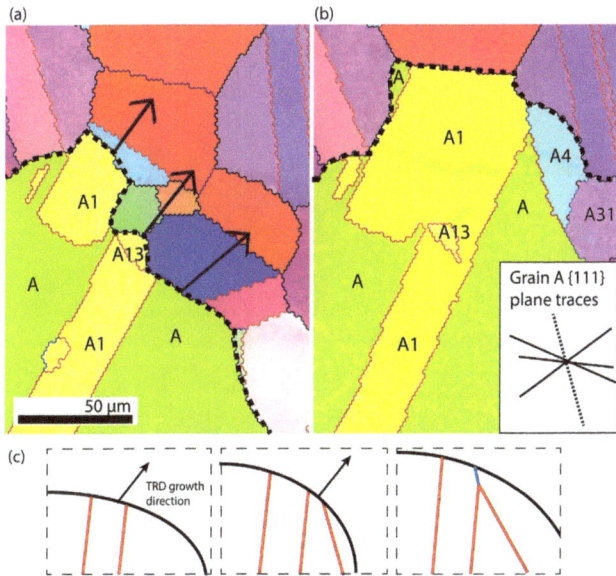

Figure 2. (**a**,**b**) Evolution of the twin cluster during the annealing process, where the dashed line indicates the interface between the existing twin cluster and the deformed matrix; (**c**) schematic of the formation of the AT boundary during the evolution of the twin cluster front and the formation of Σ9 GB (the inset illustrates the 111 plane trace for the growing grain (A). adopted from Ref. [36].

Recent studies [6,55] have shown that the deformation microstructures, including stacking faults, planar-slip dislocation structures, and deformation twins, exhibit distinctive effects on the evolution of twin clusters and thus on the GBCD optimization, as evidenced by the experimental findings of Cu-16at.% Al alloys in Figure 3. The stacking faults and planar-slip dislocation structures are fairly beneficial to the GBCD optimization, for which the formation of ATs can be induced by the ordered defects in a sequence of closely packed atomic planes at the front end of a growing twin cluster. On the contrary, deformation twins hinder the growth of twin clusters, thus impairing GBCD optimization. Therefore, the optimal prior strain for the GBCD optimization should be around the threshold strain for the appearance of deformation twins in FCC metals.

In addition, some low ΣCSL GBs can also be formed when two separate twin cluster migration fronts meet together during the TMP [36,59]. These special GBs must be located in the network of random high-angle GBs, which can effectively block the connectivity of RHAGB networks.

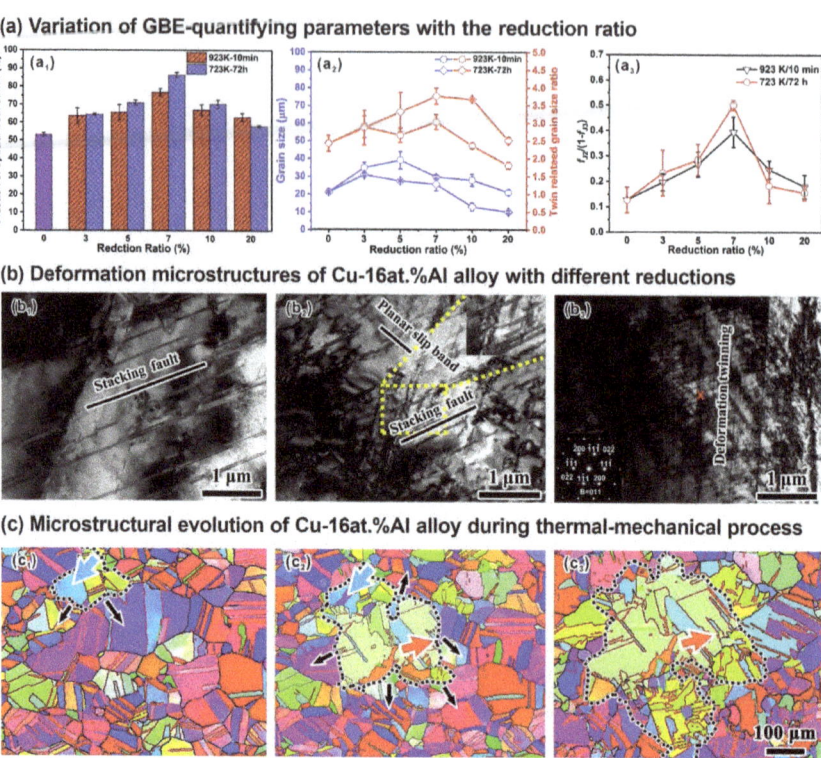

Figure 3. Influence of deformation microstructures on the GBCD optimization of the Cu-16at.%Al alloy (**a**) Variation of GBE-quantifying parameters (a_1: f_{SBs}, a_2: ratio of twin- related domain size to grain size, a_3: triple-junction distribution) with the reduction ratio; (**b**) deformation microstructures at low (3%, b_1), optimal (7%, b_2), and high (20%, b_3) reductions; (**c**) quasi-in situ EBSD observations on the evolution of microstructures during TMP treatment at an optimal stain (7% reduction, c_1) and annealing at 723 K for 12 h (c_2) and 36 h (c_3). Adopted from Ref. [6].

Figure 4 shows the schematic diagram of the "twin cluster growth" model obtained by summing up the above research results. In the model, the ATs in twin clusters are mainly induced by the planar deformation microstructures, including stacking faults and planar dislocation structures, which are the main source of special GBs. For example, once the stacking faults in deformation microstructures encounter migrating RHAGBs, can they strongly affect a sequence of close-packed planes of recrystallizing grain and induce the transformation from a regular sequence (... ABC ...) to an inverse sequence (... CBA ...), as displayed in Figure 4. This is because there is a $\frac{a}{6}[11\bar{2}]$ displacement between atoms in stacking faults and in perfect crystal. The displacement can effectively reduce the necessary energy of twinning, i.e., the stacking faults can provide excessive activation energy for twinning [36,59]. Furthermore, in a twin cluster, ATs with different orientations interact with each other as they grow with the migration of RHAGBs, thus inducing some other special GBs. Further, at the final stage of twin cluster evolution, the intersection between separate twin clusters also induces certain special GBs. Finally, the GBCD optimization of FCC metals is fully realized by the growth of twin clusters.

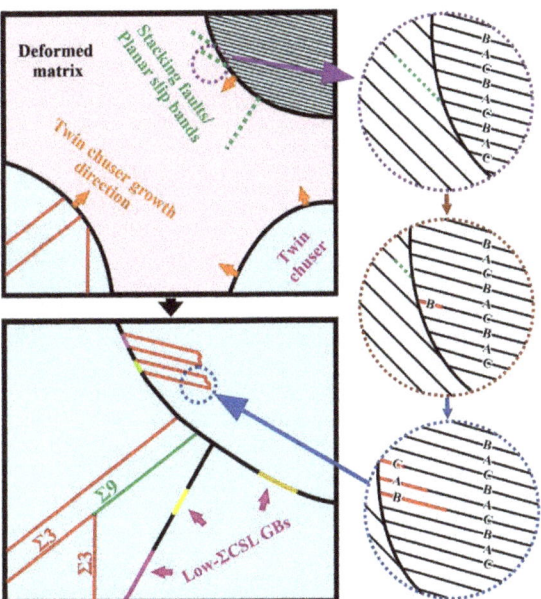

Figure 4. Schematic diagram of the "twin cluster growth" model for the GBCD optimization in FCC metals.

During the growth of twin clusters, some non-symemtric low ΣCSL GBs can also migrate in one twin cluster, such as incoherent Σ3 GBs, can reduce the interfacial energy by interacting with other special GBs in the cluster [60–62]. However, such behavior is not necessarily conducive to the increase of f_{SBs} (e.g., the disappearance of the AT boundary in Figure 2).

5. Influence of GBCD Optimization on Tensile Properties

In polycrystalline metals, the fundamental parameters for evaluating the tensile properties, such as yield strength, ultimate strength, and uniform elongation, are closely related to their microstructures. For example, the well-known fine-grain strengthening (or GB strengthening) can not only effectively improve the strength but also optimize the uniform elongation. However, there also exists a unique effect of GBCD optimization on the tensile properties of metallic materials. Further, to eliminate the influences of precipitation, phase transformation, and other special microstructures, some stable single-phase FCC metallic materials, such as austenitic stainless steel, pure copper, and copper alloys, have been selected as target materials to systematically study the influence of GBCD optimization on their tensile properties [6,12,63].

5.1. Influence of GBCD Optimization on Room-Temperature Tensile Properties

On the premise of a same-level grain size, the influence of GBCD optimization on the yield strength of single-phase FCC metals, such as austenitic stainless steel and copper alloys [6,63], can be negligible, while the influence on the ultimate strength is related to the role that different types of GBs play in the behavior of dislocation slipping. Moreover, the mode of dislocation slipping is also different in various materials due to differences in stacking fault energy, short-range order, and friction stress [44,64].

Additionally, for the FCC materials with an extremely low stacking fault energy, a perfect dislocation is very likely to dissociate into two partial dislocations (extended dislocations) that are separated by the stacking fault, and the two partial dislocations are constrained to move in a slip plane and cannot cross slip, so that their deformation behavior is generally dominated by the planar slip of dislocations and deformation twinning [65,66]. It is well known that the special GBs are high-angle GBs with low Miller indices, a small disorder degree, and low interfacial energy [51,67]. The slip planes on both sides of the interface are often continuous but with a certain turning angle, which allows some dislocations to slip across the interface. However, it must be noted that dislocations in FCC metals can slip only after the partial dislocations recombine into a perfect dislocation. According to the inverse relationship between the equilibrium width of an extended dislocation and the stacking fault energy [7], dislocations in FCC metals with an extremely low stacking fault energy are not so easy to slip across the special GBs. In this case, the functions (in terms of mechanical behavior) of special GBs are similar to those of ordinary RHAGBs. Figure 5 shows a case study regarding the influence of GBE on the room-temperature mechanical properties of Cu-16at.% Al alloy with an extremely low stacking fault energy (6 mJ/m^2). As a result, the GBE treatment has little effect on the tensile strength of the Cu-16at.% Al alloy at room temperature (Figure 5a), but it improves the ductility to a certain extent due to the fact that special GB can improve the deformation uniformity, the capacity of maintainable work hardening, and the resistance to intergranular cracking (Figure 5b) [6].

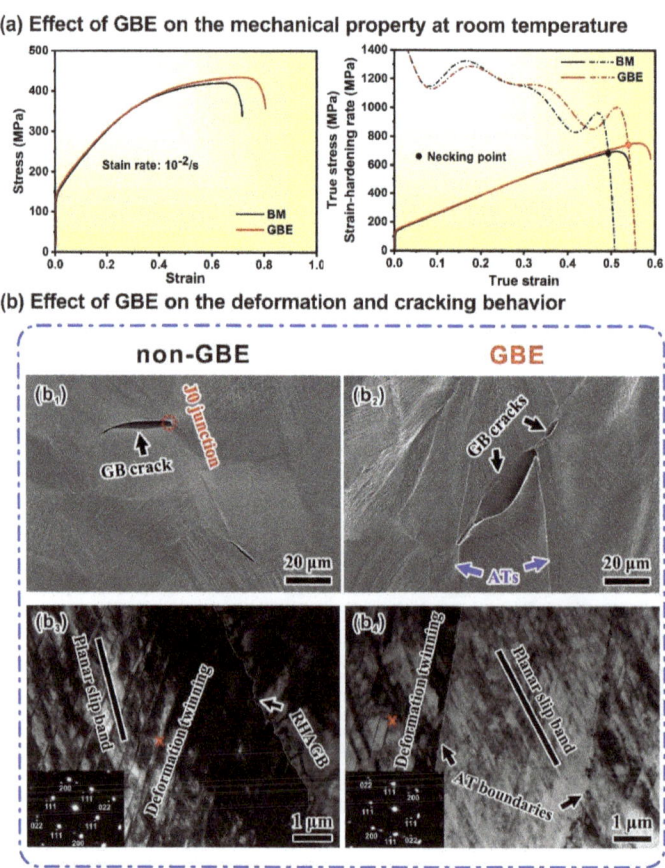

Figure 5. Effect of GBCD optimization on the room-temperature tensile properties of the Cu-16at.% Al alloy (**a**) Comparisons of the tensile properties of non-GBE and GBE samples (**b**) Influences of GBE on the deformation (b_3, b_4) and cracking (b_1, b_2) behaviors adopted from Ref. [6].

In materials with a high stacking fault energy and numerous short-range order structures, their deformation mode is still dominated by planar slip of dislocations because of the "glide plane softening" effect [68]. However, the dislocations in such materials are generally perfect dislocations or extended dislocations with a low width. As mentioned above, it is much easier for such dislocations to propagate across special GBs, which is equivalent to an increase in the average free path of dislocations [68]. According to the Kocks-Mecking model [69], the increment rate of dislocations reduces with the increase of the dislocation free path, and thereby, the work hardening behavior of these materials should be lowered by the GBCD optimization. Therefore, the ultimate tensile strength of such materials may be reduced to a certain extent after GBE treatment despite an improvement in ductility, which still needs to be further confirmed experimentally.

It is worth noting that, regardless of any materials, the destruction of the RHAGB network by special GBs should be beneficial to inhibit crack propagation, which is conducive to the improvement of ductility.

5.2. Influence of GBCD Optimization on High-Temperature Tensile Properties

Due to the continuous progress of socioeconomics, more attention has been paid to the performance of materials in some harsh environments. For instance, the mechanical performance of metallic materials at high temperatures has been particularly emphasized because it involves several important livelihood and national defense industries, such as nuclear power, aviation, and aerospace.

Previous studies [70,71] indicated that the stacking fault energy of metallic materials increased as the environmental temperature elevated, which induced the dislocation slipping mode to change from planar slip to wavy slip. Accordingly, the probability of dislocation recovery would be significantly increased, thus weakening the work-hardening capacity of materials. In addition, as the temperature reaches a high level, the dynamic recrystallization would also happen in metallic materials. Consequently, the appearance of dynamic recrystallization further increases the consumption of dislocations. Further, dislocation recovery and dynamic recrystallization are important factors causing high-temperature softening. In fact, these two unfavorable factors are closely related to the GBCD in materials. For example, due to the lower ability of RHAGBs to channel dislocation slip, the strain or stress concentration tends to happen near the ordinary RHAGBs during plastic deformation, which results in a rapid increase in the strain energy. On the one hand, the high density of dislocations gathered near RHAGBs, especially under the influence of high temperature, would cause a stronger dynamic recovery behavior, thus inducing a higher softening behavior. On the other hand, the high strain energy provides an extra driving force for the dynamic recrystallization occurring at GBs, which further aggravates the softening phenomenon. As a result, the high f_{SBs} introduced by GBE treatment can effectively weaken these two adverse effects occurring at GBs.

The reason that GBE-induced special GBs are capable of weakening the recovery of dislocations is due to the fact that the lower interfacial energy, coupled with the lower lattice distortion, will induce more moving dislocations to slip across special GBs [12], thereby reducing the local density of dislocations near the GBs. Meanwhile, GBE can reduce the Gibbs free energy of materials and thus inhibit the nucleation and growth of recrystallized grains at random high-angle GBs under high-temperature deformation, which is the major reason why GBE can weaken the dynamic recrystallization. Figure 6 shows our recent work about the influence of GBE on the mechanical properties of a Cu-16at.% Al alloy at a high temperature of 723 K and different strain rates [12]. As a result, due to the reasons above mentioned, the high-temperature strength and ductility of FCC metals (e.g., Cu-16at.%Al alloy) can be synchronously improved by a GBE treatment (Figure 6a) under the premise that dynamic recovery and recrystallization occurs more easily at a lower strain rate of 10^{-4} s^{-1} (Figure 6(b_2)), and the corresponding evidences for the suppressive effects of GBE on the GB cracking and dynamic recrystallization at a lower strain rate of 10^{-4} s^{-1}

are given, respectively, in Figure 6(c_1,c_2) (inverse pole figure maps) and Figure 6(c_3,c_4) (TEM images).

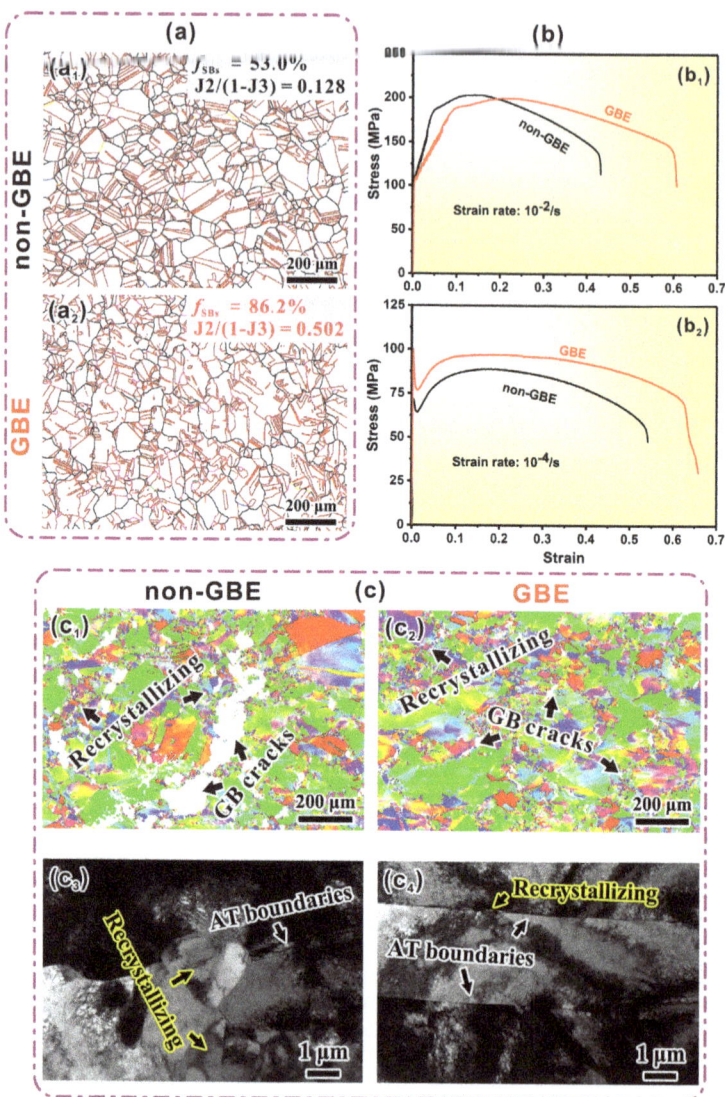

Figure 6. Influence of GBE on the tensile properties and deformation behavior of Cu-16at.% Al alloy at 723 K and at different strain rates. Comparisons of GBCD (**a**) and tensile property (**b**) between GBE and non-GBE samples, and the experimental evidence for the suppressive effects of GBE on the GB cracking (inverse pole figure maps c_1,c_2) and dynamic recrystallization (TEM images c_3, c_4). adopted from Ref. [12].

6. Influence of GBCD Optimization on Creep Properties

As the service temperature exceeds half of the melting point, the creep will generally become a crucial threat to the service safety of polycrystalline metallic materials. It is well known that the creep of metallic materials mainly originates from dislocation-climbing and GB-slipping [8]. Apparently, there is a great difference in the structural stability of various types of GBs under high-temperature deformation. In this case, the resistance of various GBs to dislocation slip should be different. Therefore, the GBE treatment is bound to have a certain impact on the creep resistance of polycrystalline metallic materials.

Firstly, the lowered Gibbs free energy in GBE samples signifies that a relatively higher activation energy for creep is needed under a high-temperature constant load. Secondly, the higher lattice order of the special GBs suppresses the dislocation decomposition at the GBs and thus significantly inhibits the GB slipping [72]. Finally, it is difficult to form cavities on the interface of special GBs by the decomposition of dislocations at high temperature, which greatly improves the resistance to creep failure in GBEed materials [72–75]. For instance, Lehockey and Palumbo [72] reported that the significant reductions in bulk primary creep strain and steady-state creep rate can be realized and the cavitation damage at GBs can be greatly suppressed in nickel by increasing the fraction of special GBs, as shown in Figure 7. In another case study, Was et al. [75] observed that the creep resistance of Ni-16Cr-9Fe alloys can also be effectively improved by the GBE treatment. In short, the GBE treatment is indeed a well-established method to improve the creep resistance of metallic materials.

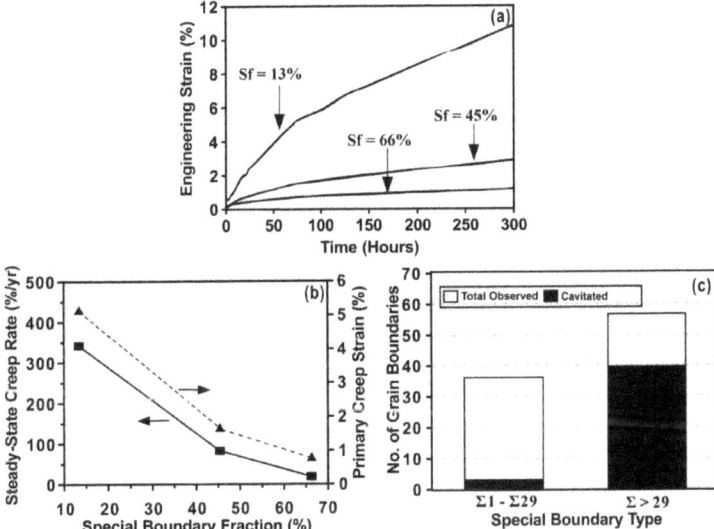

Figure 7. Influence of GBE on the creep properties of nickel. (a) Comparisons of creep curves for nickel samples with various fractions of special GBs (Sf); (b) Effect of the fraction of special GBs on the steady-state creep rate and total primary creep strain; (c) Relative proportion of RHAGBs ($\Sigma > 29$) and special GBs ($\Sigma 1$–$\Sigma 29$) showing cavitation in the sample with the Sf = 45. adopted from Ref. [72].

7. Influence of GBCD Optimization on Fatigue Performance

It is a known fact that at least half of the mechanical fracture in metal components is caused by the fatigue failure of materials [76]. In order to improve the fatigue resistance of materials, numerous studies have been carried out on the understanding of fatigue deformation behavior and fatigue cracking mode. It has been understood that the fatigue cracking of materials mainly arises from the strain localization under cyclic loading [77,78]. Moreover, the fatigue cracking mode was found to be closely dependent upon the service environment, the type of fatigue load, etc. [44,79,80]. In addition, under a low cyclic load or a relatively low temperature, fatigue cracks tend to initiate along slip bands in FCC materials; however, as the load or temperature increases, they will be more likely to form at GBs [79,81]. Furthermore, it was easier for the fatigue cracks of polycrystalline FCC materials to nucleate along slip bands under a tension-compression cyclic load than at GBs under a tensile-tension fatigue load [82]. It should be noted that, in polycrystalline metallic materials, the capacity of different types of GBs to resist fatigue cracking is also significantly different. For example, in the study of the fatigue damage behavior of copper bicrystals, Li et al. [83] indicated that, compared with ordinary RHAGBs, coherent twin boundaries have a higher resistance to fatigue cracking. Pan et al. [84] demonstrated that the cyclic stress response is independent of the loading history in nano-twin strengthened polycrystalline copper, which is beyond all doubt beneficial for the fatigue performance. Therefore, the twin-related GBE is regarded as a feasible method to improve the fatigue properties of metallic materials.

Lehockey et al. [85] explored the effect of GBE treatment on the fatigue properties of alloy 738 and alloy V-57 through tensile-tensile fatigue tests at room temperature. The high f_{SBs} introduced by GBE treatment in alloys 738 and V-57 effectively improved the fatigue lives of these two alloys. They suggested that the main reason for the GBE-induced improvement in fatigue properties should be attributed to the influences of GBE on the precipitation behavior in these two alloys [85].

In order to probe a pathway to improve the low-cycle fatigue life of FCC metals via GBE, our recent work [13] examined the tension-tension fatigue behavior of the non-GBE and GBE Cu-16at.% Al alloys at relatively high stress amplitudes, and it was found that an appropriate GBE treatment (i.e., cold rolled with 7% reduction and annealed at 723 K for 72 h) can effectively improve the stress-controlled tension-tension low-cycle fatigue life of Cu-16at.%Al alloys (Figure 8a). The GBE treatment can lead to a greater capability of compatible deformation (Figure 8b) and a higher resistance to GB cracking (Figure 8c), and thus effectively hinder the cyclic strain localization and cracking at GBs, especially at increased stress amplitudes, so that the sensitivity of fatigue life to stress amplitude can be weakened by GBE in Cu-16at.% Al alloys. This research strongly demonstrated that the GBE method can be regarded as an efficient pathway to improve low-cycle fatigue resistance of FCC metals.

In addition, through an in situ observation on the fatigue crack growth in SUS304 austenitic stainless steel, Kobayashi et al. [86,87] confirmed that the fatigue cracking resistance of special GBs is significantly better than that of RHAGBs, as shown in Figure 9. In this work, the polycrystalline specimen (Type A) with a higher fraction of low-ΣCSL boundaries shows a lower crack propagation rate compared with the specimen (Type B) with a lower fraction of low-ΣCSL boundaries (Figure 9a), since the higher fraction of low-ΣCSL boundaries significantly improves the resistance to intergranular cracking (Figure 9b).

Furthermore, the room-temperature fatigue resistance of metallic materials can be improved by a GBE treatment, which can optimize the deformation uniformity and the resistance to intergranular cracking. As mentioned above, the intergranular cracking tendency of polycrystalline metallic materials becomes more obvious at high temperatures; in this case, it will naturally raise the question of whether the effect of GBE on fatigue performance will become more significant or not. For this reason, Gao et al. [81] investigated the fatigue cracking behavior of nickel-based superalloy ME3 at high temperatures of 973 K and 1073 K, and they found that the intergranular fatigue cracking behavior in the target material indeed became significant with increasing temperature; however, it can be effectively suppressed by a GBE treatment, especially at the higher temperature of 1073 K rather than the lower temperature of 973 K. Therefore, GBE is more efficacious and helpful for improving the high-temperature fatigue resistance of metallic materials.

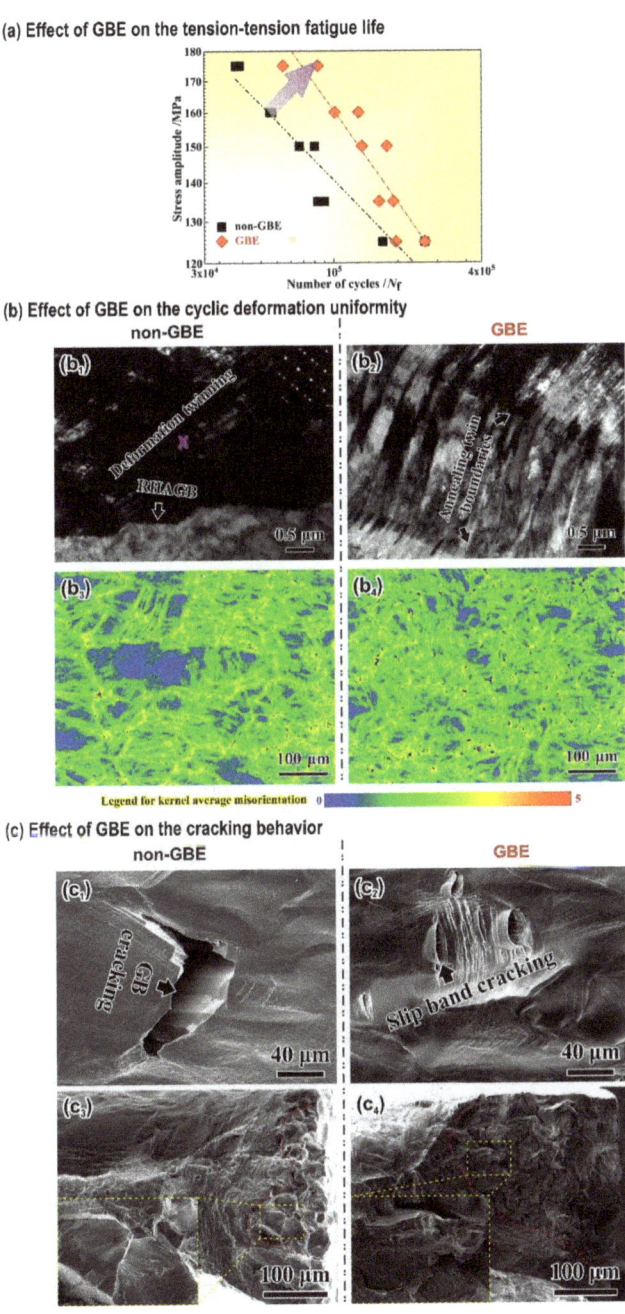

Figure 8. Effect of GBE on the Fatigue Performance of Cu-16at.% Al Alloy. (**a**) Comparison of the fatigue lives of non-GBE and GBE samples at different stress amplitudes; and comparisons of the deformation uniformity (**b**) and cracking behavior (**c**) of non-GBE and GBE samples fatigued at a stress amplitude of 175 MPa. adopted from Ref. [13].

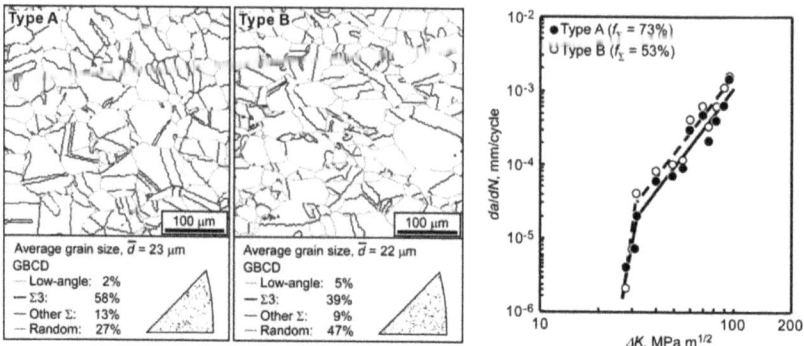

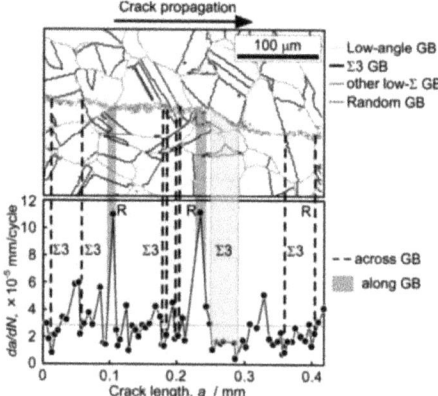

Figure 9. Shows that GBE improves the fatigue propagation resistance in SUS304 austenitic stainless steel. (**a**) Comparisons of GBCD and crack propagation rates of Type A and Type B samples (**b**) Influence of GBCD on the local propagation rate of fatigue cracks Adopted from Ref. [86].

Gao et al. [32] recently found that the GBE has little visible effect on the tension-compression fatigue property of 316LN austenitic stainless steel under a high temperature (573 K) salt solution, as shown in Figure 10a. In this work, the fatigue cracks of 316LN austenitic stainless steel mainly nucleated and propagated along slip bands (Figure 10b) due to the influence of hydrogen embrittlement, which was hardly restricted by special GBs. Accordingly, the GBE treatment may effectively improve the fatigue properties of metallic materials, for which intergranular cracking is the dominant cause of fatigue damage, but it has little effect on the fatigue damage along slip bands. Additional, in-depth work still needs to be done to elucidate the definite effect of GBE on the fatigue properties of various metallic materials.

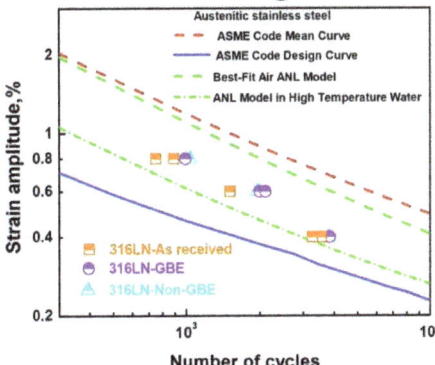

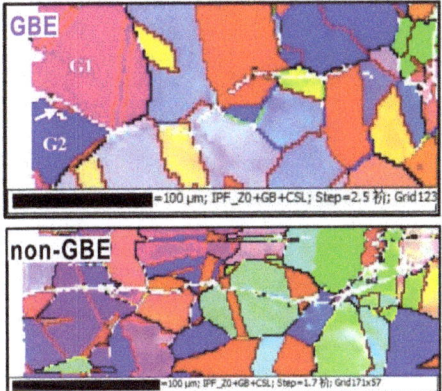

Figure 10. Effect of GBE on the fatigue performance of 316LN austenitic stainless steels in high-temperature salt solutions. (**a**) Relationship between strain amplitude and fatigue life for GBE and non-GBE samples; (**b**) Crack propagation in GBE and non-GBE samples. Adopted from Ref. [32].

8. Summary

On the basis of reaffirming the concept of GBE, this review summarizes the recent development of twin-related GBE, including the theoretical models and mechanisms of GBE optimization, and emphatically concentrates on the applications of GBE to improve the mechanical properties of polycrystalline metallic materials. The theoretical models of twin-related GBE have been relatively well developed, except for some technical problems (e.g., the determination of effective special GBE). This review has strongly demonstrated the powerful and fruitful applications of twin-related GBE in improving the various mechanical properties of polycrystalline metallic materials, e.g., tensile ductility at room temperature, strength-ductility match at high temperature, creep resistance, fatigue life, etc. Therefore, the twin-related GBE should be a feasible solution to the design of high-performance materials in the future. However, most of the existing research was just conducted on a laboratory scale, and it is still lacking in the application of practical engineering. Therefore, exploring low-cost and reliable methods to realize GBE is still a major challenge to its application. In addition, the application of GBE to optimizing the fatigue properties of FCC materials has an excellent prospect for being developed into a novel pathway, which is worthy of being further investigated.

Author Contributions: Conceptualization, X.L.; writing—review and editing, X.L., X.G.; consulting literature, Z.J., P.C., C.F., F.S. All authors have read and agreed to the published version of the manuscript.

Funding: This work was financially supported by the National Natural Science Foundation of China (NSFC) under Grant nos. 51871048 and 52171108 and the Fundamental Research Funds for the Central Universities under Grant no. N2202007.

Acknowledgments: Special thanks to the Analytical and Testing Center, Northeastern University, China.

Conflicts of Interest: The authors declare no conflict of interest.

Abbreviations

The abbreviations and their original meanings appear in this review.

GBs	Grain boundaries
RHAGBs	Random high-angle grain boundaries
GBCD	Grain boundary character distribution
GBE	Grain boundary engineering
ATs	Annealing twins
TMP	Thermal-mechanical process
FCC	Face-centered cubic
CSL	Coincidence site lattice
EBSD	Electron back-scatter diffraction
TRD	Twin related domain
TEM	Transmission electron microscope

References

1. Hondros, E.D. Reproduced optical micrograph from Sorby HC (1887). In *The Donald McLean Symposium on Structural Materials: Engineering Application through Scientific Insight*; Institute of Materials Cambridge University Press: Cambridge, UK, 1996; p. 1.
2. Watanabe, T. An approach to grain boundary design for strong and ductile polycrystals. *Res. Mech.* **1984**, *11*, 47–84.
3. Aust, K.T. Grain Boundary Engineering. *Can. Metall. Quart.* **1994**, *33*, 265–274. [CrossRef]
4. Watanabe, T. Grain boundary engineering: Historical perspective and future prospects. *J. Mater. Sci.* **2011**, *46*, 4095–4115. [CrossRef]
5. Randle, V. Grain boundary engineering: An overview after 25 years. *Mater. Sci. Technol.* **2013**, *26*, 253–261. [CrossRef]
6. Guan, X.; Shi, F.; Ji, H.; Li, X. Gain boundary character distribution optimization of Cu-16at.%Al alloy by thermomechanical process: Critical role of deformation microstructure. *Mater. Sci. Eng. A* **2019**, *765*, 138299. [CrossRef]
7. Meyers, M.A.; Chawla, K.K. *Mechanical Behavior of Materials*; Cambridge University Press: Cambridge, UK, 2009.
8. Hosford, W.F. *Mechanical Behavior of Materials*; Cambridge University Press: New York, NY, USA, 2005.
9. Qi, J.; Huang, B.; Wang, Z.; Ding, H.; Xi, J.; Fu, W. Dependence of corrosion resistance on grain boundary characteristics in a high nitrogen CrMn austenitic stainless steel. *J. Mater. Sci. Technol.* **2017**, *33*, 1621–1628. [CrossRef]
10. Shi, F.; Gao, R.-H.; Guan, X.-J.; Liu, C.-M.; Li, X.-W. Application of Grain Boundary Engineering to Improve Intergranular Corrosion Resistance in a Fe–Cr–Mn–Mo–N High-Nitrogen and Nickel-Free Austenitic Stainless Steel. *Acta Met. Sin.* **2020**, *33*, 789–798. [CrossRef]
11. Dong, X.; Li, N.; Zhou, Y.; Peng, H.; Qu, Y.; Sun, Q.; Shi, H.; Li, R.; Xu, S.; Yan, J. Grain boundary character and stress corrosion cracking behavior of Co-Cr alloy fabricated by selective laser melting. *J. Mater. Sci. Technol.* **2021**, *93*, 244–253. [CrossRef]
12. Guan, X.; Shi, F.; Ji, H.; Li, X. A possibility to synchronously improve the high-temperature strength and ductility in face-centered cubic metals through grain boundary engineering. *Scr. Mater.* **2020**, *187*, 216–220. [CrossRef]
13. Guan, X.; Jia, Z.; Liang, S.; Shi, F.; Li, X. A pathway to improve low-cycle fatigue life of face-centered cubic metals via grain boundary engineering. *J. Mater. Sci. Technol.* **2022**, *113*, 82–89. [CrossRef]
14. Shibata, N.; Oba, F.; Yamamoto, T.; Ikuhara, Y. Structure, energy and solute segregation behaviour of [110] symmetric tilt grain boundaries in yttria-stabilized cubic zirconia. *Philos. Mag.* **2004**, *84*, 2381–2415. [CrossRef]
15. Brandon, D. The structure of high-angle grain boundaries. *Acta Met.* **1966**, *14*, 1479–1484. [CrossRef]
16. Sutton, A.P.; Balluffi, R.W. *Interfaces in Crystalline Materials*; Clarendon: Oxford, UK, 1995.
17. Shimada, M.; Kokawa, H.; Wang, Z.; Sato, Y.; Karibe, I. Optimization of grain boundary character distribution for intergranular corrosion resistant 304 stainless steel by twin-induced grain boundary engineering. *Acta Mater.* **2002**, *50*, 2331–2341. [CrossRef]
18. Balluffi, R. Grain Boundary Structure and Properties. Master's Thesis, Massachusetts Institute of Technology, Cambridge, MA, USA, 1979.

19. Laws, M.; Goodhew, P. Grain boundary structure and chromium segregation in a 316 stainless steel. *Acta Met. Mater.* **1991**, *39*, 1525–1533. [CrossRef]
20. Shi, F.; Tian, P.; Jia, N.; Ye, Z.; Qi, Y.; Liu, C.; Li, X. Improving intergranular corrosion resistance in a nickel-free and manganese-bearing high-nitrogen austenitic stainless steel through grain boundary character distribution optimization. *Corros. Sci.* **2016**, *107*, 49–59. [CrossRef]
21. Michiuchi, M.; Kokawa, H.; Wang, Z.; Sato, Y.; Sakai, K. Twin-induced grain boundary engineering for 316 austenitic stainless steel. *Acta Mater.* **2006**, *54*, 5179–5184. [CrossRef]
22. Pradhan, S.; Bhuyan, P.; Mandal, S. Individual and synergistic influences of microstructural features on intergranular corrosion behavior in extra-low carbon type 304L austenitic stainless steel. *Corros. Sci.* **2018**, *139*, 319–332. [CrossRef]
23. Sinha, S.; Kim, D.-I.; Fleury, E.; Suwas, S. Effect of grain boundary engineering on the microstructure and mechanical properties of copper containing austenitic stainless steel. *Mater. Sci. Eng. A* **2015**, *626*, 175–185. [CrossRef]
24. Randle, V.; Coleman, M. A study of low-strain and medium-strain grain boundary engineering. *Acta Mater.* **2009**, *57*, 3410–3421. [CrossRef]
25. Singh, G.; Hong, S.-M.; Oh-Ishi, K.; Hono, K.; Fleury, E.; Ramamurty, U. Enhancing the high temperature plasticity of a Cu-containing austenitic stainless steel through grain boundary strengthening. *Mater. Sci. Eng. A* **2014**, *602*, 77–88. [CrossRef]
26. Randle, V.; Ralph, B. Applications of Grain Boundary Engineering to Anomalous Grain Growth. *MRS Proc.* **1988**, *122*, 419–424. [CrossRef]
27. King, W.E.; Schwartz, A.J. Toward Optimization of the Grain Boundary Character Distribution in OFE Copper. *Scr. Mater.* **1998**, *38*, 449–455. [CrossRef]
28. Tan, L.; Ren, X.; Sridharan, K.; Allen, T. Corrosion behavior of Ni-base alloys for advanced high temperature water-cooled nuclear plants. *Corros. Sci.* **2008**, *50*, 3056–3062. [CrossRef]
29. Tan, L.; Sridharan, K.; Allen, T.; Nanstad, R.; McClintock, D. Microstructure tailoring for property improvements by grain boundary engineering. *J. Nucl. Mater.* **2008**, *374*, 270–280. [CrossRef]
30. Cao, W.; Xia, S.; Bai, Q.; Zhang, W.; Zhou, B.; Li, Z.; Jiang, L. Effects of initial microstructure on the grain boundary network during grain boundary engineering in Hastelloy N alloy. *J. Alloys Compd.* **2017**, *704*, 724–733. [CrossRef]
31. Li, H.; Mao, Q.; Zhang, M.; Zhi, Y. Effects of aging temperature and grain boundary character on carbide precipitation in a highly twinned nickel-based superalloy. *Philos. Mag.* **2021**, *101*, 1274–1288. [CrossRef]
32. Gao, J.; Tan, J.; Wu, X.; Xia, S. Effect of grain boundary engineering on corrosion fatigue behavior of 316LN stainless steel in borated and lithiated high-temperature water. *Corros. Sci.* **2019**, *152*, 190–201. [CrossRef]
33. Segura, I.; Murr, L.; Terrazas, C.; Bermudez, D.; Mireles, J.; Injeti, V.; Li, K.; Yu, B.; Misra, R.; Wicker, R. Grain boundary and microstructure engineering of Inconel 690 cladding on stainless-steel 316L using electron-beam powder bed fusion additive manufacturing. *J. Mater. Sci. Technol.* **2019**, *35*, 351–367. [CrossRef]
34. Wang, W.; Yin, F.; Guo, H.; Li, H.; Zhou, B. Effects of recovery treatment after large strain on the grain boundary character distributions of subsequently cold rolled and annealed Pb–Ca–Sn–Al alloy. *Mater. Sci. Eng. A* **2008**, *491*, 199–206. [CrossRef]
35. Wang, W.; Zhou, B.; Rohrer, G.S.; Guo, H.; Cai, Z. Textures and grain boundary character distributions in a cold rolled and annealed Pb–Ca based alloy. *Mater. Sci. Eng. A* **2010**, *527*, 3695–3706. [CrossRef]
36. Barr, C.M.; Leff, A.C.; Demott, R.W.; Doherty, R.D.; Taheri, M.L. Unraveling the origin of twin related domains and grain boundary evolution during grain boundary engineering. *Acta Mater.* **2018**, *144*, 281–291. [CrossRef]
37. Sharma, N.K.; Shekhar, S. New perspectives on twinning events during strain-induced grain boundary migration (SIBM) in iteratively processed 316L stainless steel. *J. Mater. Sci.* **2021**, *56*, 792–814. [CrossRef]
38. Jin, Y.; Lin, B.; Bernacki, M.; Rohrer, A.; Bozzolo, N. Annealing twin development during recrystallization and grain growth in pure nickel. *Mater. Sci. Eng. A* **2014**, *597*, 295–303. [CrossRef]
39. Schuh, C.A.; Kumar, M.; King, W.E. Analysis of grain boundary networks and their evolution during grain boundary engineering. *Acta Mater.* **2003**, *51*, 687–700. [CrossRef]
40. Hu, H.; Zhao, M.; Rong, L. Retarding the precipitation of η phase in Fe-Ni based alloy through grain boundary engineering. *J. Mater. Sci. Technol.* **2020**, *47*, 152–161. [CrossRef]
41. Devaraj, A.; Kovarik, L.; Kautz, E.; Arey, B.; Jana, S.; Lavender, C.; Joshi, V. Grain boundary engineering to control the discontinuous precipitation in multicomponent U10Mo alloy. *Acta Mater.* **2018**, *151*, 181–190. [CrossRef]
42. Kokawa, H. Weld decay-resistant austenitic stainless steel by grain boundary engineering. *J. Mater. Sci.* **2005**, *40*, 927–932. [CrossRef]
43. Shi, F.; Yan, L.; Hu, J.; Wang, L.F.; Li, T.Z.; Li, W.; Guan, X.J.; Liu, C.M.; Li, X.W. Improving Intergranular Stress Corrosion Cracking Resistance in a Fe–18Cr–17Mn–2Mo–0.85N Austenitic Stainless Steel through Grain Boundary Character Distribution Optimization. *Acta Met. Sin.* **2022**, *35*, 1849–1861. [CrossRef]
44. Han, D.; Zhang, Y.; Li, X. A crucial impact of short-range ordering on the cyclic deformation and damage behavior of face-centered cubic alloys: A case study on Cu-Mn alloys. *Acta Mater.* **2021**, *205*, 116559. [CrossRef]
45. Irukuvarghula, S.; Hassanin, H.; Cayron, C.; Attallah, M.; Stewart, D.; Preuss, M. Evolution of grain boundary network topology in 316L austenitic stainless steel during powder hot isostatic pressing. *Acta Mater.* **2017**, *133*, 269–281. [CrossRef]
46. Kwon, Y.J.; Jung, S.P.; Lee, B.J.; Lee, C.S. Grain boundary engineering approach to improve hydrogen embrittlement resistance in FeMnC TWIP steel. *Int. J. Hydrogen Energ.* **2018**, *43*, 10129–10140. [CrossRef]

47. Lin, P.; Palumbo, G.; Erb, U.; Aust, K. Influence of grain boundary character distribution on sensitization and intergranular corrosion of alloy 600. *Scr. Met. Mater.* **1995**, *33*, 1387–1392. [CrossRef]
48. Palumbo, G.; Aust, K. Structure-dependence of intergranular corrosion in high purity nickel. *Acta Met. Mater.* **1990**, *38*, 2343–2352. [CrossRef]
49. Lehockey, E.M.; Brennenstuhl, A.M.; Palumbo, G.; Lin, P. Electrochemical noise for evaluating susceptibility of lead-acid battery electrodes to intergranular corrosion. *Br. Corros. J.* **1998**, *33*, 29–36. [CrossRef]
50. Randle, V. *The Role of the Coincidence Site Lattice in Grain Boundary Engineering*; Institute of Materials: London, UK, 1996.
51. Priester, L. *Grain Boundaries: From Theory to Engineering*; Springer Series in Materials Science; Springer Science & Business Media: Dordrecht, The Netherlands, 2013.
52. Lehockey, E.; Brennenstuhl, A.; Thompson, I. On the relationship between grain boundary connectivity, coincident site lattice boundaries, and intergranular stress corrosion cracking. *Corros. Sci.* **2004**, *46*, 2383–2404. [CrossRef]
53. Jones, R.; Randle, V. Sensitisation behaviour of grain boundary engineered austenitic stainless steel. *Mater. Sci. Eng. A* **2010**, *527*, 4275–4280. [CrossRef]
54. Kumar, M.; King, W.E.; Schwartz, A.J. Modifications to the microstructural topology in f.c.c. materials through thermomechanical processing. *Acta Mater.* **2000**, *48*, 2081–2091. [CrossRef]
55. Guan, X.; Shi, F.; Jia, Z.; Li, X. Grain boundary engineering of AL6XN super-austenitic stainless steel: Distinctive effects of planar-slip dislocations and deformation twins. *Mater. Charact.* **2020**, *170*, 110689. [CrossRef]
56. Randle, V. Mechanism of twinning-induced grain boundary engineering in low stacking-fault energy materials. *Acta Mater.* **1999**, *47*, 4187–4196. [CrossRef]
57. Kumar, M.; Schwartz, A.J.; King, W.E. Microstructural evolution during grain boundary engineering of low to medium stacking fault energy fcc materials. *Acta Mater.* **2002**, *50*, 2599–2612. [CrossRef]
58. Wang, W.; Guo, H. Effects of thermo-mechanical iterations on the grain boundary character distribution of Pb-Ca-Sn-Al alloy. *Mater. Sci. Eng. A* **2007**, *445–446*, 155–162. [CrossRef]
59. Tokita, S.; Kokawa, H.; Sato, Y.S.; Fujii, H.T. In situ EBSD observation of grain boundary character distribution evolution during thermomechanical process used for grain boundary engineering of 304 austenitic stainless steel. *Mater. Charact.* **2017**, *131*, 31–38. [CrossRef]
60. Straumal, B.B.; Kogtenkova, O.A.; Gornakova, A.S.; Sursaeva, V.G.; Baretzky, B. Review: Grain boundary faceting–roughening phenomena. *J. Mater. Sci.* **2016**, *51*, 382–404. [CrossRef]
61. Sursaeva, V.G.; Straumal, B.B.; Gornakova, A.S.; Shvindlerman, L.S.; Gottstein, G. Effect of faceting on grain boundary motion in Zn. *Acta Mater.* **2008**, *56*, 2728–2734. [CrossRef]
62. Straumal, B.B.; Polyakov, S.A.; Mittemeijer, E.J. Temperature influence on the faceting of $\Sigma 3$ and $\Sigma 9$ grain boundaries in Cu. *Acta Mater.* **2006**, *54*, 167–172. [CrossRef]
63. Zhuo, Z.; Xia, S.; Bai, Q.; Zhou, B. The effect of grain boundary character distribution on the mechanical properties at different strain rates of a 316L stainless steel. *J. Mater. Sci.* **2018**, *53*, 2844–2858. [CrossRef]
64. Li, P.; Li, S.; Wang, Z.; Zhang, Z. Fundamental factors on formation mechanism of dislocation arrangements in cyclically deformed fcc single crystals. *Prog. Mater. Sci.* **2011**, *56*, 328–377. [CrossRef]
65. Wang, Z.; Wang, Y.; Liao, X.; Zhao, Y.; Lavernia, E.; Zhu, Y.; Horita, Z.; Langdon, T. Influence of stacking fault energy on deformation mechanism and dislocation storage capacity in ultrafine-grained materials. *Scr. Mater.* **2009**, *60*, 52–55. [CrossRef]
66. Rohatgi, A.; Vecchio, K.S.; Gray, G.T. The influence of stacking fault energy on the mechanical behavior of Cu and Cu-Al alloys: Deformation twinning, work hardening, and dynamic recovery. *Metall. Mater. Trans. A* **2001**, *32*, 135–145. [CrossRef]
67. Priester, L. *Grain Boundaries and Crystalline Plasticity*; ISTE Ltd.: London, UK; John Wiley & Sons, Inc.: Hoboken, NJ, USA, 2011.
68. Han, D.; Guan, X.; Yan, Y.; Shi, F.; Li, X. Anomalous recovery of work hardening rate in Cu-Mn alloys with high stacking fault energies under uniaxial compression. *Mater. Sci. Eng. A* **2019**, *743*, 745–754. [CrossRef]
69. Mecking, H.; Kocks, U. Kinetics of flow and strain-hardening. *Acta Met.* **1981**, *29*, 1865–1875. [CrossRef]
70. Saka, H.; Sueki, Y.; Imura, T. On the intrinsic temperature dependence of the stacking-fault energy in copper-aluminium alloys. *Philos. Mag. A* **1978**, *37*, 273–289. [CrossRef]
71. Tisone, T.C.; Brittain, J.O.; Meshii, M. Stacking Faults in a Cu-15 at% Al Alloy. I. The Short Range Order and Temperature Dependence of the Stacking Fault Energy. *Phys. Status Solidi* **1968**, *27*, 185–194. [CrossRef]
72. Lehockey, E.; Palumbo, G. On the creep behaviour of grain boundary engineered nickel 1. *Mater. Sci. Eng. A* **1997**, *237*, 168–172. [CrossRef]
73. Kokawa, H.; Watanabe, T.; Karashima, S. Dissociation of lattice dislocations in coincidence boundaries. *J. Mater. Sci.* **1983**, *18*, 1183–1194. [CrossRef]
74. Yoo, M.H.; Trinkaus, H. Crack and cavity nucleation at interfaces during creep. *Met. Mater. Trans. A* **1983**, *14*, 547–561. [CrossRef]
75. Was, G.S.; Thaveeprungsriporn, V.; Crawford, D.C. Grain boundary misorientation effects on creep and cracking in Ni-based alloys. *JOM* **1998**, *50*, 44–49. [CrossRef]
76. Suresh, S. *Fatigue of Materials*; Cambridge University Press: Cambridge, UK, 1998.
77. Differt, K.; Esmann, U.; Mughrabi, H. A model of extrusions and intrusions in fatigued metals II. Surface roughening by random irreversible slip. *Philos. Mag. A* **1986**, *54*, 237–258. [CrossRef]

78. Chen, Y.; Cao, Y.; Qi, Z.; Chen, G. Increasing high-temperature fatigue resistance of polysynthetic twinned TiAl single crystal by plastic strain delocalization. *J. Mater. Sci. Technol.* **2021**, *93*, 53–59. [CrossRef]
79. Shao, C.W.; Shi, F.; Li, X.W. Cyclic deformation behavior of Fe-18Cr-18Mn-0.63 N nickel-free high-nitrogen austenitic stainless steel. *Metall. Mater. Trans. A* **2015**, *46*, 1610–1620. [CrossRef]
80. Tan, J.; Wu, X.; Han, E.-H.; Ke, W.; Liu, X.; Meng, F.; Xu, X. Corrosion fatigue behavior of Alloy 690 steam generator tube in borated and lithiated high temperature water. *Corros. Sci.* **2014**, *89*, 203–213. [CrossRef]
81. Gao, Y.; Ritchie, R.O.; Kumar, M.; Nalla, R.K. High-cycle fatigue of nickel-based superalloy ME3 at ambient and elevated temperatures: Role of grain-boundary engineering. *Met. Mater. Trans. A* **2005**, *36*, 3325–3333. [CrossRef]
82. Kim, G.-H.; Kwon, I.-B.; Fine, M.E. The influence of loading methods on fatigue crack initiation in polycrystalline copper at ambient temperature. *Mater. Sci. Eng. A* **1991**, *142*, 177–182. [CrossRef]
83. Li, L.; Zhang, Z.; Zhang, P. Higher fatigue cracking resistance of twin boundaries than grain boundaries in Cu bicrystals. *Scr. Mater.* **2011**, *65*, 505–508. [CrossRef]
84. Pan, Q.; Zhou, H.; Lu, Q.; Gao, H.; Lu, L. History-independent cyclic response of nanotwinned metals. *Nature* **2017**, *551*, 214–217. [CrossRef]
85. Lehockey, E.M.; Palumbo, G.; Lin, P. Improving the weldability and service performance of nickel-and iron-based superalloys by grain boundary engineering. *Met. Mater. Trans. A* **1998**, *29*, 3069–3079. [CrossRef]
86. Kobayashi, S.; Hirata, M.; Tsurekawa, S.; Watanabe, T. Grain boundary engineering for control of fatigue crack propagation in austenitic stainless steel. *Procedia Eng.* **2011**, *10*, 112–117. [CrossRef]
87. Kobayashi, S.; Nakamura, M.; Tsurekawa, S.; Watanabe, T. Effect of grain boundary microstructure on fatigue crack propagation in austenitic stainless steel. *J. Mater. Sci.* **2011**, *46*, 4254–4260. [CrossRef]

Disclaimer/Publisher's Note: The statements, opinions and data contained in all publications are solely those of the individual author(s) and contributor(s) and not of MDPI and/or the editor(s). MDPI and/or the editor(s) disclaim responsibility for any injury to people or property resulting from any ideas, methods, instructions or products referred to in the content.

Article

Evolution of Poisson's Ratio in the Tension Process of Low-Carbon Hot-Rolled Steel with Discontinuous Yielding

Hai Qiu * and Tadanobu Inoue

Research Center for Structural Materials, National Institute for Materials Science, 1-2-1 Sengen, Tsukuba 305-0047, Japan
* Correspondence: qiu.hai@nims.go.jp

Abstract: Low-carbon hot-rolled steel generally undergoes a deformation process composed of four phases, i.e., elastic deformation, discontinuous yielding, work hardening, and macroscopic plastic-strain localization in a tension test. The evolution of the Poisson's ratio in terms of the average Poisson's ratio and the local Poisson's ratio in the deformation process from the non-load state to the onset point of specimen necking was investigated. The main results are as follows: (1) the average Poisson's ratio cannot accurately represent the local Poisson's ratio in the discontinuous-yielding phase; (2) the Poisson's ratio varied significantly within a plastic band in the discontinuous-yielding phase, and the maximum Poisson's ratio was reached within the plastic band; and (3) the strain rate greatly increased the Poisson's ratio.

Keywords: low-carbon steel; Poisson's ratio; digital image correlation; plastic deformation; discontinuous yielding; strain rate

1. Introduction

Materials respond to stress by straining. Under an applied stress, a material deforms (expansion or contraction) in directions parallel and perpendicular to the direction of the applied stress, i.e., resulting in strains along the two directions. The negative of the ratio of the later strain to the former strain is defined as the Poisson's ratio ($\nu = -\frac{strain\ perpendicular\ to\ the\ applied\ stress}{strain\ parallel\ to\ the\ apllied\ stress}$) [1]. Material characteristic (anisotropy of microstructure, strain state of material) and loading conditions (strain rate, temperature) influence the ν. For instance, when the strain state of a structural steel changes from the elastic state to the plastic state in a tension process, its Poisson's ratio greatly increases [2,3]. If the microstructure of the steel is isotropic, in this change process of strain sate, the transition of the ν is from a low value (0.27 to 0.3) [2,4] to an upper limit value (0.5) [2]. When the microstructure of an elastic media is anisotropic and the rotation of microscopic clusters in it occurs, the upper limit of the ν could exceed 0.5 [5]. The ν in the elastic state is usually a material constant [2–4], but it is a function of the applied strain in the large plastic strain regime [3,6]. The influence of the strain rate on the ν depends on the material types: an enhanced strain rate decreases the ν of porous titanium [7] and polymeric foams [8] but increases that of polyoxymethylene [9]. The strain rate hardly affects the ν of shale [10]. Test temperature is another factor influencing the ν. Raising the test temperature increases the ν of 8–18 stainless steel [11].

Structural steels are widely used in engineering structures, such as bridges, ships, oil tanks, and so on. To ensure the integrity of structures, steels are commonly designed to be in service in the elastic strain regime [12]. Therefore, as an elastic constant, the ν is mainly concerned with the elastic deformation region, and only the ν in the elastic regime is given in data handbooks [4,11]. However, a database on the evolution of the ν in a whole deformation process is important in some engineering applications, as in the case of a reinforced concrete (RC) structure. Low-carbon steel bars are generally embedded in the concrete to reinforce the strength of the RC structure. Bond slip between the concrete and the steel bar causes nonlinearity of the RC structure. The ratio of the lateral deformation to the longitudinal deformation of the reinforced steel bar, i.e., the ν significantly affects the bond behavior.

Although low-carbon steel is a widely used structural material, the evolution of its ν in a tension process has been little investigated. Low-carbon steel usually has two typical deformation processes. Figure 1 illustrates the two deformation processes from the no-load state to final fracture in a tension test via the stress–strain curve, in which a specimen is tensioned along the longitudinal direction. The whole deformation process, as shown in Figure 1a, is classified into four phases: Phase 1, the elastic deformation phase; Phase 2, the discontinuous-yielding phase; Phase 3, the work hardening phase; and Phase 4, the specimen-necking phase (macroscopic plastic-strain localization). Deformation over the gauge length of the specimen is uniform only in Phase 1 and Phase 3. The discontinuous-yielding phase is a non-uniform deformation region in which one plastic band first forms and then propagates across the unyielded specimen at a high speed. In the Phase 4, after the onset of specimen necking, deformation mainly concentrates in the necked region and the parts of the specimen beyond the necked region hardly deform. In contrast to the deformation process with discontinuous yielding, the deformation process with continuous yielding (round stress–strain curve), as shown in Figure 1b, is simpler—the deformation is uniform before the onset of specimen necking.

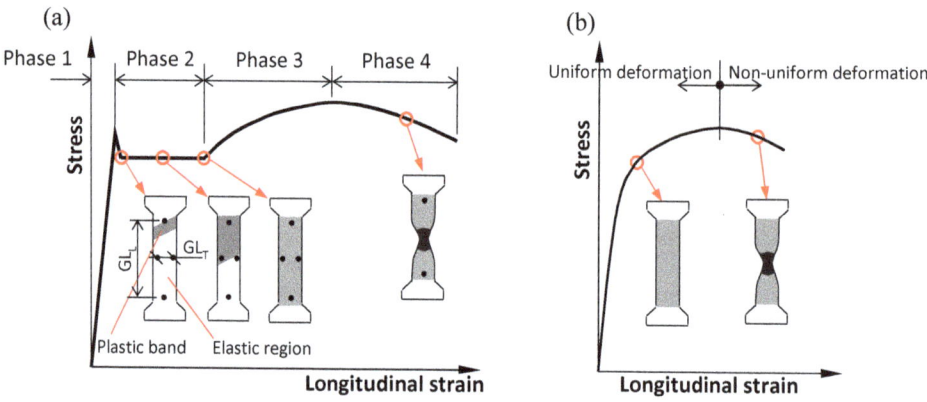

Figure 1. Illustration of two typical deformation processes via the stress–strain curve with discontinuous yielding (**a**) and with continuous yielding (**b**).

In the definition of the ν, the two strains are produced by the same applied stress instead of multiple stresses. Therefore, the ν must be experimentally measured under the uniaxial loading. The ν is determined by $\nu = -\varepsilon_T/\varepsilon_L$ in a uniaxial tension test where ε_T and ε_L are the transversal strain and the longitudinal strain, respectively. If average strains (or local strains) are used, the obtained ν is the average Poisson's ratio (or local Poisson's ratio). Conventionally, an extensometer and a strain gauge are used to measure the average strain. Recently, a digital image correlation (DIC) technique has been verified to be an effective tool to measure the average and local strains [3,13,14], and the Poisson's ratio can be determined using this technique [2,15]. If the deformation over the specimen is uniform, the distribution of the Poisson's ratio is uniform and the average Poisson's ratio is identical to the local Poisson's ratio. In contrast to the uniform deformation, non-uniform deformation will induce a difference between the two parameters. A non-uniform deformation phase is present in both the deformation processes, as shown in Figure 1. Deformation with discontinuous yielding is more complicated than that with continuous yielding, and thus the present study is concerned only with the former deformation (i.e., the stress–strain curve) in Figure 1a.

Eiriksson et al. [3] investigated the evolution of the ν of a steel bar with discontinuous yielding (i.e., the type in Figure 1a) in a tension test via the DIC technique. In their study, the average longitudinal strain over the longitudinal gauge length (GL_L) and the average transversal strain over the transversal gauge length (GL_T) (the GL_L and the GL_T are shown in Figure 1a) were used. Therefore, the obtained ν is the average ν instead of the local ν. As for the evolution of the local ν, relevant reports have not been found. In the study of Eiriksson et al., unsolved questions remain:

(1) In Figure 1a, the onset point of specimen necking is a key point. On a macroscopic scale, the specimen is in a uniaxial stress state before the point, while the necked part is in a multiaxial stress state after the point. The longitudinal and transversal strains in the necked part were produced by the longitudinal and transversal stresses. According to the definition of the ν, the coefficient of the transverse strain to the longitudinal strain is not the ν. This point was not mentioned in their study.

(2) In Phase 2, deformation is non-uniform, and thus the average ν cannot accurately represent the local value in some regions. The valid range of the average ν and the error between the average ν and the local ν need to be investigated.

(3) A plastic band is a highly localized plastic-strain region, and it is a feature of discontinuous yielding. Its correlation with the ν was unknown.

(4) The factors affecting the ν should be revealed.

The data on the evolution of the ν in a deformation process from the elastic strain state to the large plastic strain state is basic data, that can be used in some applications, such as the simulation of deformation processes. In the present study, we tried to obtain this basic data for low-carbon hot-rolled steel with discontinuous yielding. The three questions aforementioned ((2) to (4)) were discussed through uniaxial tension tests performed at room temperature, in which the strains were measured by an extensometer, strain gauges, and the DIC technique. The main research points are as follows: (1) The evolutions of the average and local ν were determined. The valid range of the average ν and its error relative to the local ν in the three deformation phases (Phases 1–3) were investigated. (2) The distribution of the ν within a plastic band in the discontinuous-yielding phase was measured. (3) Strain rate is a factor affecting the ν. Its correlation with the ν was revealed.

2. Materials and Methods

Commercial hot-rolled steel (SM490 steel) was used. Its chemical composition is 0.16% C, 1.46% Mn, and 0.44% Si. Its microstructure is composed of ferrite and pearlite. Dog-bone-type specimens were machined from a SM490 plate. The specimen size is shown in Figure 2a. As shown in Figure 2b, the front surface was sprayed with white and black paint to make speckles for DIC analysis, and two strain gauges (grid area: 1 mm × 1.1 mm) and an extensometer with a gauge length of 30 mm were attached to the back surface. These specimens were tensioned along the longitudinal direction (the x-axis) at room temperature and at a crosshead speed of 0.01 mm/s. The deformation process on the front surface within an area of 30 mm × 8 mm was continuously recorded with a digital camera at a time interval of 0.5 s. The digital images obtained were processed using VIC-2D software with a subset size of 9 pixel × 9 pixel (246 μm × 246 μm) and a step of 5 pixels (137 μm). In the DIC operation, the displacement uncertainty is 0.02 pixels.

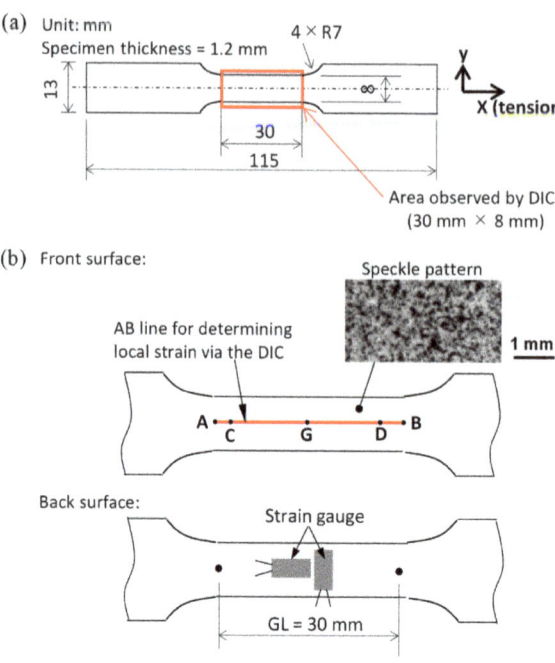

Figure 2. (**a**) Specimen size; (**b**) front surface covered by a speckle pattern. The AB line (center line) was used in DIC data processing to determine the strain. Two strain gauges were used to measure the strains along the x-axis and the y-axis on the back surface. An extensometer was attached to the back surface along the longitudinal center line to measure the longitudinal strain.

Two strain gauges, an extensometer, and the DIC technique were used to measure the strains. The obtained strains are summarized as follows: (1) The longitudinal and transversal strains obtained via strain gauges on the back surface are the average strains over the area (strain gauge grid) of 1 mm × 1.1 mm. (2) The longitudinal strain on the back surface obtained via an extensometer is the average strain over the gauge length of 30 mm. (3) As shown in Figure 2b, an AB line (center line) on the front surface was drawn for the DIC data processing. Its length is almost equal to the gauge length of the extensometer used. The average values of the longitudinal and transversal strains over the AB line were taken as the average strains. Local longitudinal and transversal strains at any point on the front surface were also determined via DIC.

3. Results and Discussion
3.1. Longitudinal and Transversal Stress–Strain Curves

At an applied stress, strains were induced along the longitudinal and transversal directions. The relation of the longitudinal (or transversal) strain to the applied stress in the whole tension process was expressed via the longitudinal (or transversal) stress–strain curve. The induced strains were measured using an extensometer, strain gauges, and DIC. The corresponding stress–strain curves and the validity of the three strain measurement methods were discussed in this section.

An extensometer is a conventional tool used to measure the strain of a bulk material. It generally captures the whole response of a bulk material from zero load to complete fracture. However, its accuracy in the elastic region is not high. In contrast to the extensometer, a strain gauge can accurately measure a small strain, and it is a reliable tool to measure the deformation behavior within the elastic region. The DIC technique is relatively new, and it is usually used to measure a large strain. The longitudinal stress–strain curves, in which the average longitudinal strain ($\varepsilon_{av.x}$) was measured via the three tools, are shown in Figure 3.

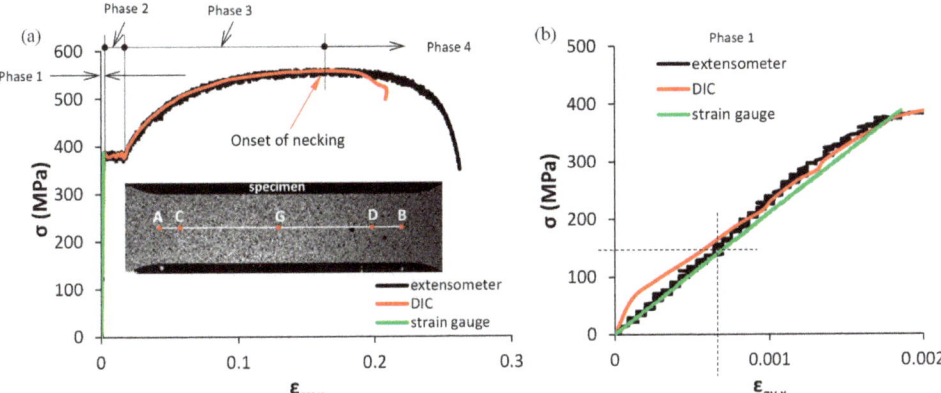

Figure 3. (**a**) The curve of the stress (σ) versus the longitudinal average strain ($\varepsilon_{av.x}$), and (**b**) the enlargement of (**a**) within Phase 1 (elastic region). The $\varepsilon_{av.x}$ was measured using an extensometer with a gauge length of 30 mm, DIC (over the AB line), and a strain gauge.

Figure 3a shows the whole deformation process from Phase 1 to Phase 4. To show the accuracy of the strain measurement, the stress–strain curve within the elastic region is enlarged in Figure 3b. It can be seen from Figure 3b that DIC has a low accuracy for a small strain, which agrees with the experimental results reported in the literature [3]. When the applied stress exceeds about 166 MPa, the difference in the elastic longitudinal strain among the three tools is very small. The strain gauge was only used in the elastic region. Figure 3a shows that DIC has almost the same accuracy as the extensometer until the occurrence of severe specimen necking.

A strain gauge and DIC were used to measure the average transversal strains. Their transversal stress–strain curves are shown in Figure 4. Apparently, the transversal strain is almost half that of the longitudinal strain. As shown in Figure 4b, DIC has low accuracy below a stress level of 301 MPa (much higher than the 166 MPa for the longitudinal strain). Therefore, the DIC data below a stress level of 301 MPa were not used to determine the Poisson's ratio in later sections.

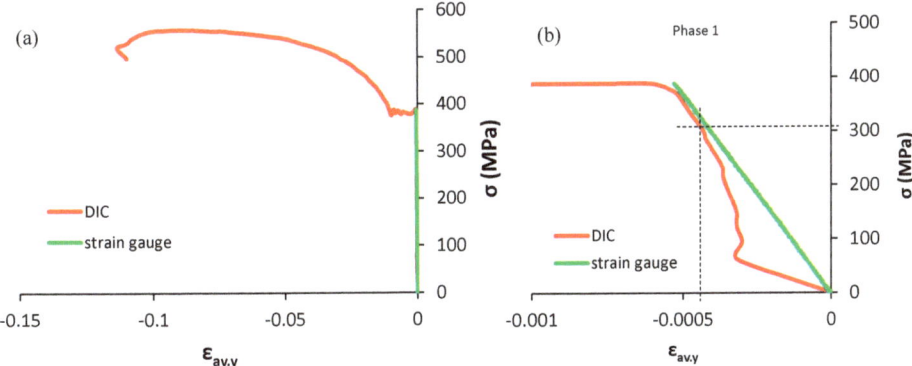

Figure 4. (**a**) The curve of the stress (σ) versus the transversal average strain ($\varepsilon_{av.y}$), and (**b**) the enlargement of (**a**) within Phase 1 (elastic region). The $\varepsilon_{av.y}$ was measured using DIC (over the AB line) and a strain gauge.

3.2. Poisson's Ratio

This section focuses on two points:

(1) The evolution of the ν in the tension process

The ν was quantitatively evaluated in terms of the average ν and the local ν, and the evolution of the ν is discussed from global and local viewpoints.

(2) Interpretation of the characteristic of the ν

In the evolution of the ν, some features were exhibited, for example, the peak curve of the ν in the discontinuous-yielding region. The correlation of those features with the strain rate is investigated.

3.2.1. The Evolution of the Average ν in the Tension Process

The average longitudinal strain ($\varepsilon_{av.x}$) and the average transversal strain ($\varepsilon_{av.y}$) obtained via the strain gauges within Phase 1 (i.e., the elastic region) are shown in Figure 5. Fitting the data gives the following formula:

$$\varepsilon_{av.y} = -0.2818\varepsilon_{av.x} \tag{1}$$

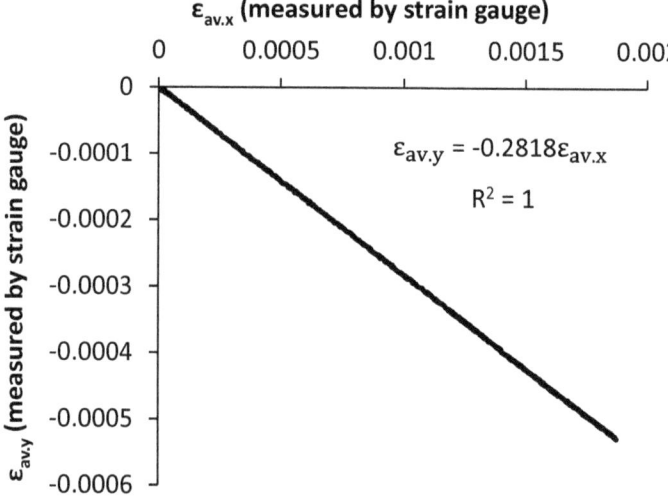

Figure 5. The longitudinal average strain ($\varepsilon_{av.x}$) versus the transversal average strain ($\varepsilon_{av.y}$) within Phase 1. The strains were measured using strain gauges.

The fitting coefficient R is equal to 1. Equation (1) shows that the average Poisson's ratio ($\nu_{av} = -\varepsilon_{av.y}/\varepsilon_{av.x}$) in the elastic region is 0.282. The value of R indicates that the strain gauge measurement has excellent accuracy to measure the average Poisson's ratio in the elastic region.

It is known that the shoulder of the tension specimen affects the deformation behavior around it. The AB line (cf. Figures 2b and 3a) was used to determine the average value of the longitudinal and transversal strains in the DIC data processing. It should be clarified which part of the AB line was influenced by the specimen shoulder. It is recognized that the parallel part of the tension specimen should be uniformly deformed within Phase 1 and Phase 3, i.e., the local strain along the AB line should be uniform. In Figure 6, two points (W1 in Phase 3, and E1 in Phase 1) on the stress–strain curves are selected. The corresponding local longitudinal strains ($\varepsilon_{loc.x}$) obtained via DIC along the AB line are shown in Figure 6. The local strain is almost uniform for a certain length, but both outside edges deviate because of the influence of the shoulder. If the influence of the shoulder is involved, a great error will be induced. Therefore, the CD line was taken as the effective length instead of the AB line to determine the average Poisson's ratio ($v_{av.CD}$). Its length is 22.9 mm (about 80% of the length of the AB line). It is noted that the AB and CD lines in Figure 6 correspond to those in Figure 3a.

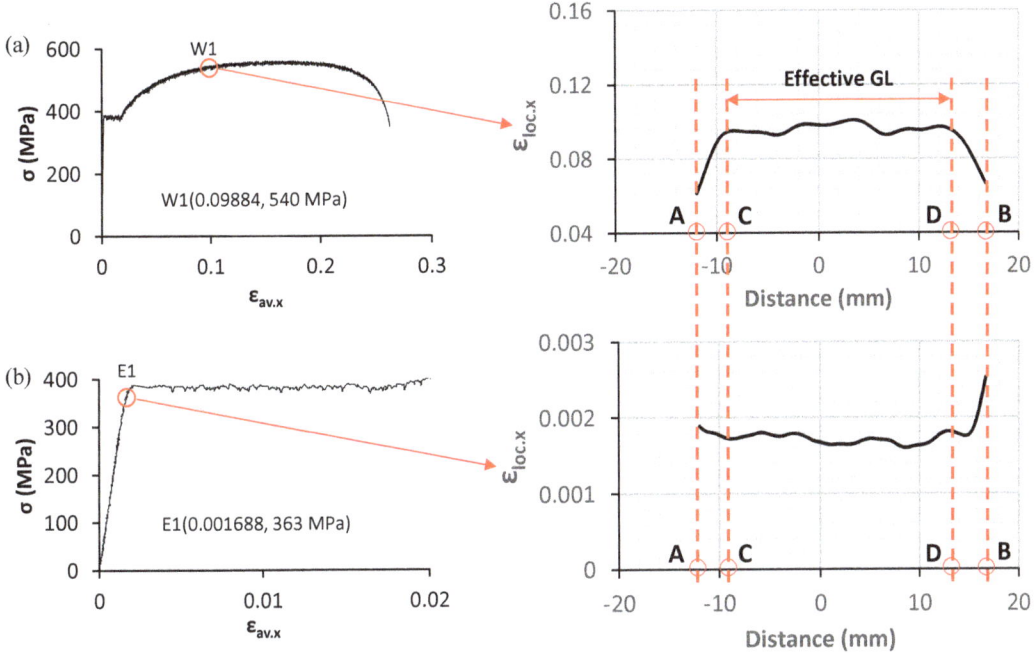

Figure 6. The longitudinal local strain ($\varepsilon_{loc.x}$) distribution along the AB line shown in Figure 3a at (a) point W1 in Phase 3, and (b) point E1 in Phase 1. The effective gauge length (GL) (i.e., the length of the CD line) shown in Figure 3a is 22.9 mm.

The average values of the local longitudinal strain ($\varepsilon_{loc.x}$) and the local transversal strain ($\varepsilon_{loc.y}$) over the CD line were taken as the average longitudinal strain ($\varepsilon_{av.x.CD}$) and the average transversal strain ($\varepsilon_{av.y.CD}$), respectively. The average Poisson's ratio ($v_{av.CD}$) was given by $-\varepsilon_{av.y.CD}/\varepsilon_{av.x.CD}$. The evolution of the $v_{av.CD}$ in the deformation process is shown in Figure 7a. Since the deformation process is conventionally expressed by the stress–strain curve, the longitudinal stress–strain curve obtained with the extensometer is plotted simultaneously. To clearly show the detail of the $v_{av.CD}$, enlarged curves are shown in Figure 7b–d.

As shown in Figure 7b, in the elastic region (Phase 1), only the $v_{av.CD}$ data corresponding to stress levels higher than 301 MPa were adopted. The $v_{av.CD}$ varies in the range of 0.26–0.31, and its average value is 0.297. The average Poisson's ratio in the elastic region determined by the strain gauge is 0.282. The average Poisson's ratios obtained using the two methods are almost the same, and their values agree with the data reported in the literature [2,4]. In Phase 2, the $v_{av.CD}$ monotonously increases as the tension increases and reaches the maximum value (0.597) at the end of Phase 2 (cf. Figure 7c). Figure 7d shows that as the deformation enters into Phase 3, the $v_{av.CD}$ first decreases and then remains almost the same, even if slight specimen necking takes place. As the specimen necking becomes severe, the $v_{av.CD}$ decreases.

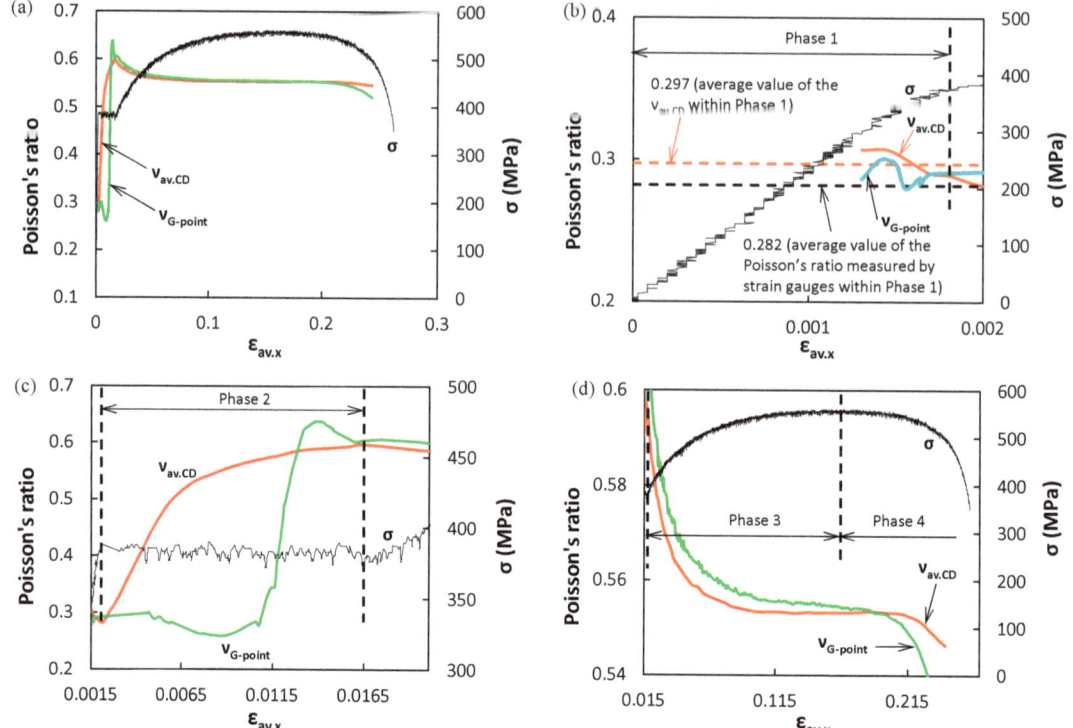

Figure 7. (a) The evolution of the average Poisson's ratio over the CD line ($\nu_{av \cdot CD}$) and the local Poisson's ratio at point G, as shown in Figure 3a ($\nu_{G\text{-point}}$), in the tension process (from Phase 1 to Phase 4). The tension process is expressed by the stress–strain curve (σ versus $\varepsilon_{av.x}$). (b) The enlargement of (a) within Phase 1. (c) The enlargement of (a) within Phase 2. (d) The enlargement of (a) within Phases 3 and 4.

3.2.2. The Evolution of the Local Poisson's Ratio in the Tension Process

As shown in Figure 3a, point G is on the AB line. The post-tested specimen shows that point G was in the necked region of the specimen. This means that point G experienced a complete process from the uniaxial stress state (before the onset of specimen necking) to biaxial stress state (after the onset of specimen necking). Therefore, the local Poisson's ratio at point G ($\nu_{G-point}$) was used to show the evolution of the local Poisson's ratio in the tension process.

The evolution of the $\nu_{G-point}$ in the whole tension process is shown in Figure 7. The $\nu_{G-point}$ shows a trend similar to that of the $\nu_{av.CD}$, but significant differences between the average and local Poisson's ratios in some regions are seen, for example, in Phase 2 (cf. Figure 7c).

It is believed that deformation in Phase 1 and Phase 3 is almost uniform. In Figure 6, point W1 and point E1 are in Phase 3 and Phase 1, respectively. The local Poisson's ratio along the CD line ($\nu_{loc.CD}$) at the two stress levels is shown in Figure 8. The average Poisson's ratio and the SD (standard deviation) are also given in Figure 8. The SD values indicate that the error at point W1 is small, and that at point E1 is allowable. Therefore, it is rational to use the average Poisson's ratio ($\nu_{av.CD}$) to describe the local Poisson's ratio.

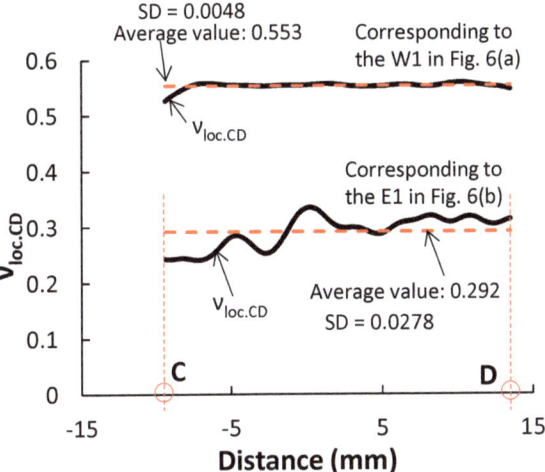

Figure 8. The distribution of the local Poisson's ratio ($v_{loc.CD}$) along the CD line corresponding to point W1 and point E1 in Figure 6. SD, standard deviation.

Phase 2 involves the whole discontinuous-yielding process. This means that the plastic region and the elastic region exist simultaneously except at the beginning and ending points of Phase 2. In Figure 9a, two points (D1, D2) are selected. The maps of local strain rate along the x-axis ($\dot{\varepsilon}_{loc.x}$), the local longitudinal strain ($\varepsilon_{loc.x}$), and the local transversal strain ($\varepsilon_{loc.y}$) corresponding to points D1 and D2 are shown in Figure 9b and c, respectively. Although band-like regions appear in the three types of maps, only the bands in the $\dot{\varepsilon}_{loc.x}$ map represent the moving plastic bands [16,17]. Two plastic bands were formed. Band-1 propagates from left to right, and Band-2 goes in the opposite direction. The two bands made contact at point D2 and finally coalesced and completely disappeared at the end of Phase 2. This indicates that two plastic regions began to occur at both outside regions, extended toward the central zone (still in the elastic state), and finally merged.

The AB line and the CD line are shown in the $\dot{\varepsilon}_{loc.x}$ map. The distributions of the $\varepsilon_{loc.x}$ and the $\varepsilon_{loc.y}$ along the CD line extracted from the corresponding maps are plotted in Figure 9b,c. The local Poisson's ratio at a point on the CD line was given by $v_{loc.CD} = -\varepsilon_{loc.y}/\varepsilon_{loc.x}$, and the distribution of the $v_{loc.CD}$ along the CD line is also given in Figure 9b,c. The plastic band width (BW) can be identified in the $\dot{\varepsilon}_{loc.x}$ map. The positions of points C and D and the band width in each figure are indicated by the vertical dotted lines. In contrast to the two moving bands in Figure 9b, which have not completely merged with the CD line, they completely merged with the CD line in Figure 9c. The widths of the two moving bands at the CD line are clearly shown in Figure 9c. It can be seen that the local longitudinal and local transversal strains ($\varepsilon_{loc.x}$, $\varepsilon_{loc.y}$) and the local Poisson's ratio ($v_{loc.CD}$) change significantly within the bandwidth—the $\varepsilon_{loc.x}$ and $\varepsilon_{loc.y}$ increase from an elastic strain to a very high plastic strain, and the $v_{loc.CD}$ from 0.282 to 0.640. The Poisson's ratio in the plastic region is significantly larger than that in the elastic region. The great difference in the Poisson's ratio between the elastic region and the plastic region leads to the great error between the average Poisson's ratio and the local Poisson's ratio.

The experimental results obtained in this section indicate that in the elastic deformation regime and the work-hardening regime, the average Poisson's ratio is identical to the local Poisson's ratio, and thus the data of the Poisson's ratio in the two regimes are basic data and can be directly used in some applications, such as calculation or simulation of the deformation process of the structural steel. However, more attention must be paid on the discontinuous-yielding regime. The elastic region and plastic region to the discontinuous-yielding regime must be separately treated by using local value instead of average value.

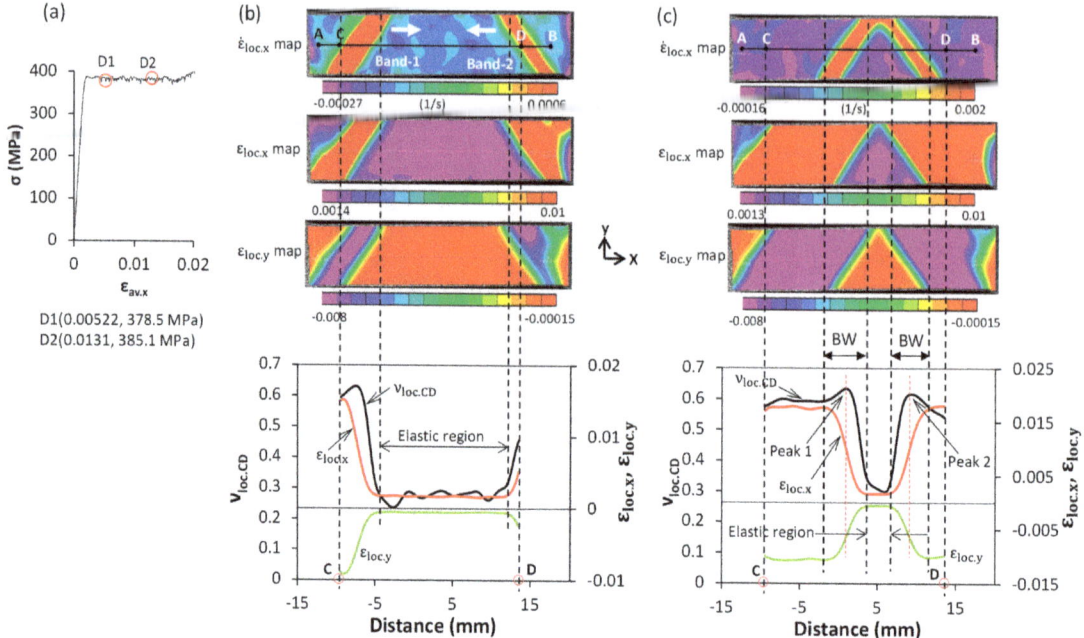

Figure 9. (a) Two points of interest (D1, D2) on the stress–strain curve in Phase 2. (**b**,**c**) Maps of the local strain rate along the x-axis ($\dot{\varepsilon}_{loc.x}$), the longitudinal local strain ($\varepsilon_{loc.x}$), and the transversal local strain ($\varepsilon_{loc.y}$). The $\varepsilon_{loc.x}$, $\varepsilon_{loc.y}$, and $\nu_{loc.CD}$ data were extracted along the CD line from the corresponding maps; (b,c) correspond to points D1 and D2, respectively. BW, bandwidth.

3.2.3. Correlation of the Poisson's Ratio with the Strain Rate

Gere and Timoshenko [2] reported that the Poisson's ratio of steel varies within the range of 0.27 to 0.5 from the elastic state to the complete plastic state—in the elastic region, it is normally in the range of 0.27 to 0.30; 0.5 is a theoretical upper limit for the Poisson's ratio of a plastic phase on the assumption that the volume of the media is constant. The present study shows that the Poisson's ratio of the steel used is about 0.282 in the elastic region, which agrees with the data in the literature [2,4]. Its Poisson's ratio in the plastic region is greater than 0.5. The steel used is commercial hot-rolled steel, and anisotropy of the microstructure is present. This probably makes the Poisson's ratio of plastic phases higher than 0.5. Wojciechowski [18] established a 2D model showing that a crystalline or polycrystalline 2D system could have a Poisson's ratio greater than 0.5.

Figure 7 shows the evolution of the average Poisson's ratio ($\nu_{av.CD}$) over the CD line and the local Poisson's ratio ($\nu_{G-point}$) at point G in the whole tension process. When point G enters into the plastic state from the elastic state, its $\nu_{G-point}$ naturally rapidly increases. This trend agrees with the conventional understanding of Poisson's ratio. In contrast to the previous results [3], in which the Poisson's ratio changes slightly in the completely plastic region, a peak exists in Phase 2 (cf. Figure 7). A similar peak phenomenon was observed in the evolution of the $\nu_{av.CD}$. This phenomenon was not reported in the literature. Eiriksson et al. [3] investigated the Poisson's ratio of a similar steel via DIC, but they did not find this peak phenomenon. Round bar specimens were used in their study. Because of the curvature of the specimen, the focus along the transversal direction varies, decreasing the accuracy of the speckle patterns on the digital images along the transversal direction. As a result, the error of the transversal strain obtained from these images via DIC was large. Eiriksson et al. [3] reported that the Poisson's ratio of steel in the elastic region was 0.37, which deviates considerably from the range of 0.27 to 0.30. This experimental data verified the low measurement accuracy for the transversal strain. For this reason, their studies did not detect the peak phenomenon of the Poisson's ratio.

The distribution of the local Poisson's ratio ($v_{loc.CD}$) along the CD line is shown in Figure 9c. Two peaks (Peak 1 and Peak 2) are present. Two vertical dotted red lines are drawn from the two peak points. The cross-points of the two straight lines with the curves of the $\varepsilon_{loc.x}$ and the $\varepsilon_{loc.y}$ are the local longitudinal strain and the local transversal strain, respectively, corresponding to the two peaks. The corresponding local strains are almost half of the maximum local plastic strains. The conventional understanding is that as the local plastic strain increases, the local Poisson's ratio increases. However, the peak phenomenon did not occur at the maximum local plastic strain, i.e., it did not agree with this conventional understanding. There must be a particular factor leading to this inconsistency.

The tension tests were performed at a constant speed, and thus it seemed that the strain rate over the specimen was almost the same in the tension process before the onset of specimen necking. However, it has been reported that when discontinuous yielding is present, the strain rate over the specimen is not uniform, and the strain rate within the moving band is much higher (almost one order) than the strain rate of the bulk material [16]. As mentioned previously, the peak phenomenon of the Poisson's ratio is one feature of the discontinuous-yield phase (Phase 2). The correlation of this phenomenon with the strain rate is shown in Figure 10. The evolution of the local Poisson's ratio ($v_{G-point}$) and the local strain rate along the longitudinal direction ($\dot{\varepsilon}_{G-point.x}$) at point G during the tension process are shown in Figure 10a. In the whole tension process, the curve of the local strain rate has two peaks. The local strain-rate map showed that after the formation of a plastic band in Phase 2, the band propagated across the specimen. It is known that the plastic band is a highly strain-rate-localized region, and the maximum local strain rate within the plastic band is about ten times that in the neighboring region. The movement of the plastic band seems like the movement of a highly localized region of high strain rate, and thus when the band crossed point G, it inevitably produced a peak in the local strain rate. Point G was in the necked-specimen region. After the occurrence of specimen necking, deformation was mainly concentrated in the necked region, and other parts were hardly deformed. As a result, the localized deformation led to the second peak of the local strain rate. The first peak of the local strain rate corresponds to the peak point of the local Poisson's ratio. The peak of the local Poisson's ratio is apparently attributed to the high local strain rate. However, a similar correlation of the high strain rate with the local Poisson's ratio is not seen with the second peak of the local strain rate. The two different effects of the strain rate were caused by the stress state of the specimen. For the first peak of the local strain rate, point G was in the uniaxial stress state; however, for the second peak of the local strain rate, the stress state of point G turned into a biaxial stress state due to the specimen necking. Specimen necking increased the restraint of the deformation along the transversal direction, and this effect weakened the effect of the strain rate. It is noted that because the range of the $v_{G-point}$ corresponding to the second peak of the local strain rate is in the necked region, the values of ($v_{G-point}$) within the range do not represent the Poisson's ratio.

Figure 9b,c shows that at a given stress point in Phase 2, peaks of the local Poisson's ratio appear at certain points along the CD line. The correlation of the peak phenomenon with the strain rate is shown in Figure 10b. The fact that the peaks of the local Poisson's ratio correspond to the peaks of strain rate indicates that a high strain rate led to the peaks of the local Poisson's ratio. Peak 1 and Peak 2 are indicated by two arrows in Figure 10b. The local Poisson's ratio for Peak 1 is larger than that of Peak 2. The strain rate corresponding to Peak 1 is greater than that for Peak 2. This means that the higher the strain rate is, the larger the Poisson's ratio is.

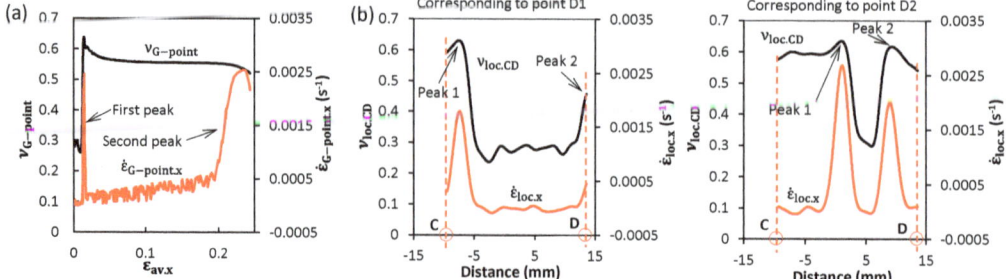

Figure 10. Correlation of the Poisson's ratio with the strain rate. (**a**) The evolution of the local Poisson's ratio ($\nu_{G-point}$) and the longitudinal strain rate ($\dot{\varepsilon}_{G-point.x}$) at point G in the tension process. (**b**) The distributions of the local Poisson's ratio ($\nu_{loc.CD}$) and the longitudinal local strain rate ($\dot{\varepsilon}_{loc.x}$) along the CD line at two stress points (D1 and D2) are shown in Figure 9a.

4. Conclusions

The evolution of the Poisson's ratio of a low-carbon hot-rolled steel in a tension process was measured using strain gauges and a digital image correlation technique. The Poisson's ratio was evaluated in terms of the average Poisson's ratio, which was the average value over a gauge length, and the local Poisson's ratio at a given point. The following main results were obtained:

(1) The distribution of the Poisson's ratio was generally uniform in the elastic deformation regime and the hard-working regime, but it was non-uniform in the discontinuous-yielding regime. The average Poisson's ratio was almost identical to the local Poisson's ratio in the elastic deformation regime and the hard-working regime. In the discontinuous-yielding regime, the average Poisson's ratio cannot accurately express the local Poisson's ratio.

(2) The values of the Poisson's ratio were obtained as follows: 0.28 in the elastic regime; 0.28 to 0.64 in the discontinuous-yielding regime; and 0.55 to 0.59 in the completely plastic deformation regime.

(3) The Poisson's ratio changed significantly within a moving plastic band in the discontinuous-yielding phase. A high strain rate within the plastic band enhanced the Poisson's ratio. The maximum local strain rate within the band induced the maximum local Poisson's ratio.

Author Contributions: Conceptualization, H.Q. and T.I.; methodology, H.Q.; software, H.Q.; validation, H.Q. and T.I.; formal analysis, H.Q.; investigation, H.Q.; resources, H.Q.; data curation, H.Q. and T.I.; writing—original draft preparation, H.Q.; writing—review and editing, H.Q. and T.I.; visualization, H.Q.; supervision, H.Q. All authors have read and agreed to the published version of the manuscript.

Funding: This research received no external funding.

Data Availability Statement: Data sharing is not applicable to this article.

Conflicts of Interest: The authors declare no conflict of interest.

References

1. Ashby, M.F.; Jones, D.R.H. *Engineering Materials 1, An Introduction to Properties, Applications and Design*, 3rd ed.; Elsevier Ltd: London, UK, 2005; p. 35.
2. Gere, J.M.; Timoshenko, S.P. *Mechanics of Materials*, 3rd ed.; Chapman & Hall: London, UK, 1991.
3. Eiriksson, H.J.; Bessason, B.; Unnthorsson, R. Uniaxial and lateral strain behavior of ribbed reinforcement bars inspected with digital image correlation. *Struct. Concr.* **2018**, *19*, 1992–2003. [CrossRef]
4. JIMM (The Japan Institute of Materials and Metals), ISIJ (The Iron and Steel Institute of Japan). *Handbook of Steels*; Maruzen: Tokyo, Japan, 1967.

5. Dmitriev, S.V.; Shigenari, T.; Abe, K. Poisson ratio beyond the limits of the elasticity theory. *J. Phys. Soc. Jpn.* **2001**, *70*, 1431–1432. [CrossRef]
6. Zimin, B.A.; Smirnov, I.V.; Sudenkov, Y.V. Behavior of lateral-deformation coefficients during elastoplastic deformation of metals. *Mechanics* **2017**, *62*, 306–309. [CrossRef]
7. Wang, B.R.; Sun, T.; Fezzaa, K.; Huang, J.Y.; Luo, S.N. Rate-dependent deformation and Poisson's effect in porous titanium. *Mater. Letters* **2019**, *245*, 134–137. [CrossRef]
8. Bhagavathula, K.B.; Meredith, C.S.; Ouellet, S.; Romanyk, D.L.; Hogan, J.D. Density, strain rate and strain effects on mechanical property evolution in polymeric foams. *Int. J. Impact Eng.* **2022**, *161*, 104100. [CrossRef]
9. Filanova, Y.; Hauptmann, J.; Längler, F.; Naumenko, K. Inelastic behavior of polyoxymethylene for wide strain rate and temperature ranges: Constitutive modeling and identification. *Materials* **2021**, *14*, 3667. [CrossRef]
10. Wei, Y.L.; Zao, L.Y.; Yuan, T.; Liu, W. Study on mechanical properties of shale under different loading rates. *Front. Earth Sci.* **2022**, *9*, 815616. [CrossRef]
11. ISIJ (The Iron and Steel Institute of Japan). *Handbook of Steels, I. Fundamentals*, 3rd ed.; Maruzen: Tokyo, Japan, 1981.
12. Anderson, T.L. *Fracture Mechanics, Fundamentals and Applications*, 3rd ed.; Taylor & Francis: London, UK, 2005.
13. Peters, W.H.; Ranson, W.F. Digital imaging techniques in experimental stress analysis. *Opt. Eng.* **1982**, *21*, 427–431. [CrossRef]
14. Sutton, M.A.; Wolters, W.J.; Peters, W.H.; Ranson, W.F.; McNeill, S.R. Determination of displacements using and improved digital correlation method. *Image Vis. Comput.* **1983**, *1*, 133–139. [CrossRef]
15. Bai, R.X.; Jiang, H.; Lei, Z.K.; Liu, D.; Chen, Y.; Yan, C.; Wang, T.; Chu, Q.L. Virtual field method for identifying elastic-plastic constitutive parameters of aluminum alloy laser welding considering kinematic hardening. *Opt. Lasers Eng.* **2018**, *110*, 122–131. [CrossRef]
16. Qiu, H.; Inoue, T.; Ueji, R. Experimental measurement of the variables of Lüders deformation in hot-rolled steel via digital image correlation. *Mater. Sci. Eng. A* **2020**, *790*, 139756. [CrossRef]
17. Qiu, H.; Inoue, T.; Ueji, R. In-situ observation of Lüders band formation in hot-rolled steel via digital image correlation. *Metals* **2020**, *10*, 530. [CrossRef]
18. Wojciechowski, K.W. Remarks on "Poisson ratio beyond the limits of the elasticity theory". *J. Phys. Soc. Jpn.* **2003**, *72*, 1819–1820. [CrossRef]

Disclaimer/Publisher's Note: The statements, opinions and data contained in all publications are solely those of the individual author(s) and contributor(s) and not of MDPI and/or the editor(s). MDPI and/or the editor(s) disclaim responsibility for any injury to people or property resulting from any ideas, methods, instructions or products referred to in the content.

Article

Repair Reliability Analysis of a Special-Shaped Epoxy Steel Sleeve for Low-Strength Tee Pipes

Jun Cao [1,*], Haidong Jia [2], Weifeng Ma [1], Ke Wang [1,3], Tian Yao [1], Junjie Ren [1], Hailiang Nie [1], Xiaobin Liang [1] and Wei Dang [1]

1. State Key Laboratory of Performance and Structural Safety for Petroleum Tubular Goods and Equipment Materials, CNPC Tubular Goods Research Institute, Xi'an 710077, China
2. Pipeline Network Group (Xinjiang) United Pipeline Co., Ltd., Urumqi 830012, China
3. School of Materials Science and Engineering, Xi'an University of Technology, Xi'an 710048, China
* Correspondence: caojun1@cnpc.com.cn; Tel.: +86-29-8818-7903

Abstract: Ensuring the safe operation of pipe fittings in a natural gas station is critical. The irregular shape of the tee easily leads to uneven mechanical properties in the manufacturing process. The strength of the tee may be lower than the requirements due to its unqualified heat-treatment process. As a result, selecting a reliable way of repairing low-strength tee pipes is a pressing concern. To repair the low-strength tee pipes, a special-shaped epoxy steel sleeve (SSESS) was designed. To optimize the critical design parameters, the SSESS design criteria were established. Following that, the SSESS repair testing was conducted using the optimized design parameters. The SSESS repair reliability was proved using hydraulic burst testing with strain monitoring and simulations of unrepaired and SSESS repaired tees. The result indicated that the SSESS repaired tee's yielding and burst pressure increased, demonstrating its repair reliability. Furthermore, the SSESS repair revealed the stress and strain concentration decrease law.

Keywords: special-shaped epoxy steel sleeve; low-strength tee; strain monitoring; stress concentration

1. Introduction

The irregular shape of a large diameter tee pipe is an important aspect of oil and gas transmission pipeline engineering, but it can easily cause unequal performance during the production process. The mechanical characteristics of the large-diameter tee in batches might not fully meet the technical criteria during the heat treatment process of the tee due to numerous causes, such as the irregularity of the tee structure and the non-strict implementation of the heat treatment procedure [1]. Some unqualified tees are not screened out by performance sampling due to the latency of performance non-destructive testing procedures. In addition, irregularly shaped tees inherently suffer from local plastic deformation during the pressure-bearing process [2]. As a result, an issue has been proposed: how to cope with in-service unqualified strength tee.

If a type-B sleeve [3,4] or a special-shaped type-B sleeve [5], is used in a natural gas station, pressure reduction and welding are necessary. However, shutting down the station and dealing with the on-site welding process is challenging. Although the composite repair is simple [6], there are issues with the tee's abdominal winding, which is insufficient for mending the tee's unique shape. The composite's aging problem cannot be remedied [7,8], and the low-strength tee cannot be repaired permanently. In addition, a solution involving a steel hose junction was used for the high-pressure hoses and junctions to overcome the external damage [9].

Epoxy sleeve repair (ESR) is distributed by PII [10]. Two steel shells with a slightly larger diameter than the pipeline to be repaired are connected to cover the damaged part of the pipeline. The sleeve is installed on the surface of the pipeline in the field, and the two ends are sealed, and then the epoxy resin is injected to fill the pores between the

pipeline and the repaired sleeve. The epoxy grout compound forms an excellent bond at both steel interfaces, providing a high reinforcement of the damaged section, in the axial and circumferential directions. British Gas developed a variation of the sleeve repair concept in the form of their epoxy-filled shell repair [11]. Wood [12] elaborated on the key advantages of ESR in detail. The steel compression sleeves are introduced in the PRCI [4] and CSA Z662 [13], which is similar to the ESR. The defects of the welded joint were repaired by ESR, and proved by numerical simulation [14]. Mazurkiewicz et al. [15] studied the repair reliability of fiber glass sleeves using the burst test, and numerical simulation, and verify selected sleeve thickness. Arif et al. [16] optimized the thickness of the repair sleeve using the failure pressure estimation and simulation of sleeve installation pressure, in 24 cases ranging from 6 to 60 inches, and the optimization approach should approximate the conditions of the tests performed. Recently, Jaszak et al. [17] proposed a methodology of leakage prediction in gasketed flange joints at deformation based on finite element methods by applying a complex and multi-stage method. Li et al. [18] applied a special steel sleeve filled with epoxy resin for the leakage of the welding connection platform on the brine pipeline. However, the special-shaped ESR for the tee was rarely reported.

In this study, a special-shaped epoxy steel sleeve (SSESS) method was proposed to address the low-strength in-service tee pipe. The purpose of this study is to propose an approach to improve safety factor and eliminate potential safety hazards. The SSESS method has two advantages: nonstop transmission and high reliability. The design criteria of SSESS were presented to optimize the critical design parameters. Then, the SSESS test was performed based on the optimization design parameters. Hydraulic burst tests with strain monitoring and simulations of unrepaired and SSESS tees were carried out to prove the repair reliability of the SSESS. According to comparisons of the strain monitoring, simulation, and hydraulic burst curves between unrepaired and SSESS tees, the reliability of the SSESS repaired low-strength tee was verified.

2. Materials and Methods

2.1. Materials Properties

A 1000 mm × 1000 mm × 800 mm tee was detected as unqualified strength of X70 steel by indentation, metallurgic replica, and Leeb hardness technologies. However, the critical tensile properties of the tee were obtained from the tee of the same batch and furnace. The tensile properties of the low-strength tee are listed in Table 1.

Table 1. The properties of the tee of the same batch and furnace.

	Yield Strength (MPa)	Ultimate Tensile Strength (MPa)	Elongation (%)
S1 (branch pipe)	343	539	29.5
S2 (shoulder)	374	502	29.5
S3 (main pipe)	363	553	27.5

2.2. Design of the Special-Shaped Steel Sleeve

The repair process of low strength tee mainly included the special-shaped steel sleeve and the epoxy resin material. The special-shaped steel sleeve is of interest in this study. The special-shaped steel sleeve structure adopted the welding structure. The stress analysis is the theoretical basis for designing the special-shaped steel sleeve. The strength theory's equivalent stress or stress strength is the failure criterion. Some researchers have also used progressive failure analysis to identify the failure locations in the simulation of composite structures under aeroelastic loading [19]. The design calculation of the steel shell special-shaped epoxy sleeve under the condition of pipeline reinforcement was mainly based on the standard of GB50251-2015 (gas pipeline engineering design). One of the critical parameters of a special-shaped sleeve design was to calculate the wall thickness, which was considered under the designing pressure of 12 MPa.

According to GB50251-2015, the calculation formula for sleeve minimum thickness δ can be given as:

$$\delta = \frac{PD}{2\sigma_s \varphi Ft} \quad (1)$$

where P is the designing pressure, D is the outer diameter of the pipe, σ_s is the yield strength, φ is the welding parameter, F is the strength design coefficient, t is the temperature reduction coefficient. In this study, a tee with a specification of 1000 mm × 1000 mm × 800 mm was chosen as the test object. In addition, the material of the tee used 16 Mn of its wide applications and low cost. φ and t were set as 1, respectively, and F was set as 0.4. Hence, the minimum δ of the main and branch pipes were calculated as ~44 mm and ~36 mm, respectively. So, the design δ values of the main and branch pipes should be larger than 44 mm and 36 mm, respectively.

Another critical parameter of the SSESS is the length between the edge of the sleeve and the girth weld, as shown in Figure 1. Due to the unequal wall thickness girth weld between the tee and straight pipe, stress concentration is easily formed in the root toe of the girth weld. Therefore, six SSESS repair modes were simulated to repair the low strength 1000 mm × 1000 mm × 800 mm tee. The repair modes of six kinds of SSESS mainly changed the length of L_h as −175 mm, −100 mm, 0 mm, 100 mm, 200 mm, and 300 mm, respectively, as shown in Figure 2. According to the simulation results, the yielding pressure could be obtained in every case. The maximum allowance working pressure (MAWP) was calculated based on the strength design coefficient of 0.5, as shown in Figure 3. As shown in Figure 3, the length of L_h needs to be larger than 0. The special-shaped sleeve must contain the girth weld to the MAWP larger than the designing pressure of 12 MPa. The MAWP tends to be constant when L_h is larger than 100 mm. So, the design value of L_h needs to be larger than 100 mm. The design and material grade of equal-length studs and hexagon nuts were based on the standards of GB/T 901 (Equal-length studs) and GB/T 6170 (Hexagon nuts) in China, respectively. The material grade of equal-length studs and hexagon nuts were 10.9 and 10, respectively.

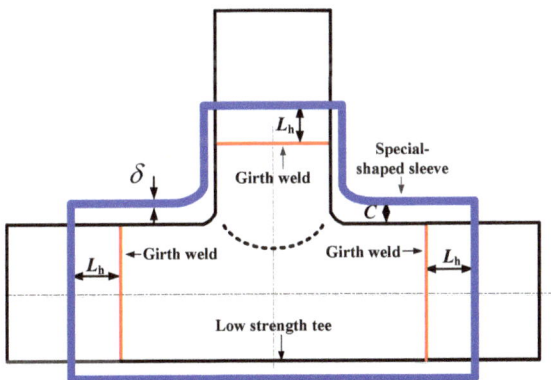

Figure 1. Critical design parameters of the special-shaped steel sleeve.

The gap C between the sleeve and tee was set as 25~40 mm, based on the fluidity and solidification time. In addition, the design of the injection and exhaust holes for the special-shaped sleeve should follow a principle, the injection and the exhaust holes need to be designed on the top and bottom of the sleeve, respectively, as shown in Figure 4. The shape and size of the design of an SSESS for 1000 mm × 1000 mm × 800 mm tee is shown in Figure 5. The holes of the positioning bolt need to be designed on the edges of the sleeve, as shown in Figure 5. The bolt fastening is used for the connection of the sleeve, as shown in Figure 5.

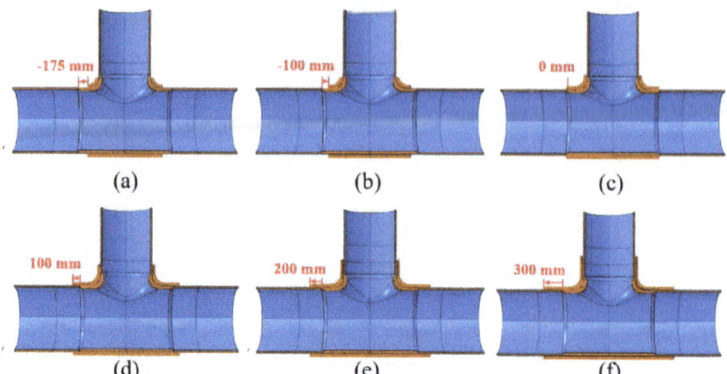

Figure 2. Six numerical-simulation verified SSESS repair modes with different L_h: (**a**) −175 mm, (**b**) −100 mm, (**c**) 0 mm, (**d**) 100 mm, (**e**) 200 mm, (**f**) 300 mm.

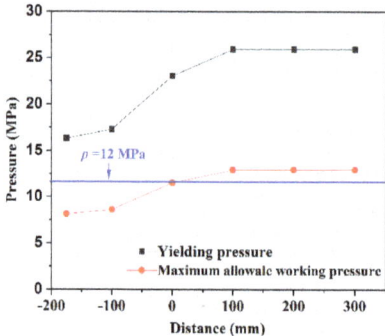

Figure 3. Yielding pressure and MAWP with the change of L_h.

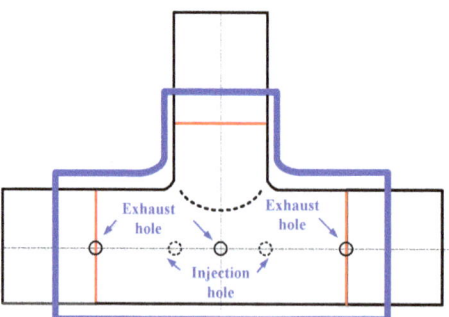

Figure 4. Design of exhaust and injection holes for special-shaped steel sleeves.

2.3. Repairing Test of SSESS

The hydraulic test structural part containing the 1000 mm × 1000 mm × 800 mm tee was welded for repair and hydraulic tests to verify the reliability of the SSESS. The procedures of repair and hydraulic tests were introduced as follows:

Step 1: Surface cleaning and sand-blasting

The repairing areas of the outer surface of the hydraulic test structural part and the inner surface of the special-shaped sleeve need to be cleaned and sandblasted until an Sa2.5 grade of requirement is reached, as shown in Figure 6.

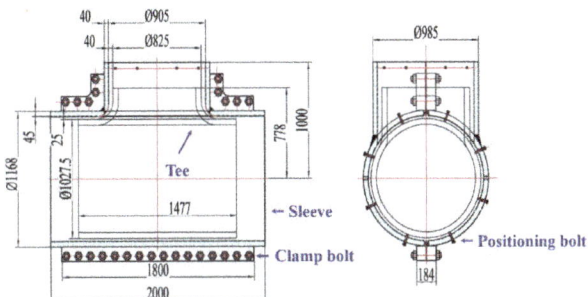

Figure 5. Design drawing of the special-shaped steel sleeve.

Figure 6. Surface cleaning and sand-blasting: (**a**) low-strength tee, and (**b**) special-shaped sleeve.

Step 2: Experiments with strain monitoring

To monitor the strain concentration and repairing effect, external strain gauges were pasted on the outer surface of the special-shaped sleeve. The external strain gauge locations and actual paste process are shown in Figure 7.

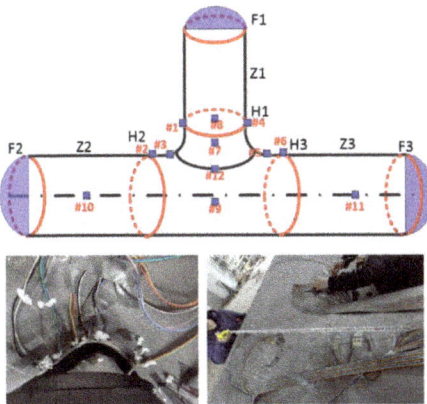

Figure 7. Bonding process of the strain gauge. The locations of these strain gauges were designed to study the critical regions: the girth weld region (#1, #2, #3, #4, #7, #8), shoulder (#3 and #5), belly (#12), the bottom of the main pipe (#9), and the reference area (#10 and #11).

Step 3: Installation of a special-shaped steel sleeve

The installation of a special shaped sleeve was carried out by crane and aided by testers. The main procedures of installation are positioning, gap adjustment, and bolt fastening, as shown in Figure 8. The critical points of installation are as follows: (I) the installation process cannot damage the strain gauge; (II) the gap between the tee and special-shaped sleeve needs to be uniformly adjusted; (III) the gasket should be used in the bolt fastening process.

Figure 8. Installation process of special-shaped steel sleeves: (**a**) positioning, (**b**) gap adjustment, and (**c**) bolt fastening.

Step 4: Injecting process

The rubber sealing tape was used to seal three ends of the special-shaped sleeve, as shown in Figure 9a. The A and B components of the injection were mixed according to a certain mass ratio (100:30) as epoxy resin, and the high-pressure airless injection pump was used to inject the resin. The epoxy resin adhesive needs to meet the requirements listed in Table 2. The injection material was injected from the two injection holes of the special-shaped sleeve through the high-pressure airless injection pump, as shown in Figure 9b. The three exhaust holes are designed on the top of the sleeve, which effectively exhausts the air between the sleeve and the tee, and makes the tee, epoxy resin, and special-shaped sleeve perfectly integrated. Once the epoxy resin is filled in the gap between the tee and sleeve, the empty bottles will timely exhibit the black epoxy resin, and the injecting process should be stopped immediately, as shown in Figure 9c.

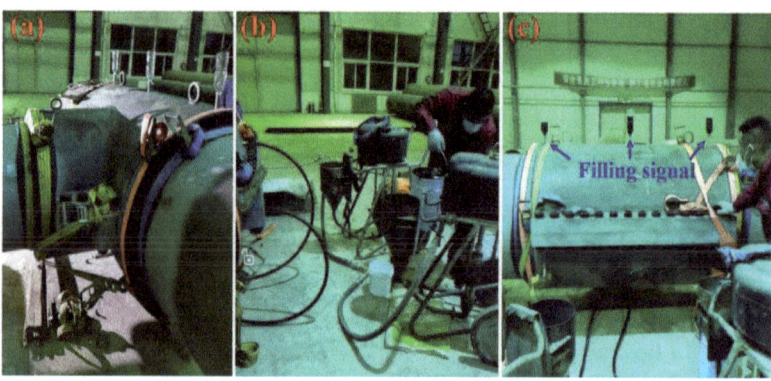

Figure 9. Injecting process of SSESS: (**a**) end sealing, (**b**) injection resin, and (**c**) filling signal.

Table 2. Properties of epoxy resin.

Properties	Unit	Requirements
Hardness of the resin after curing	Shore D	80 ± 10
Mass solid content	%	≥99.5
Curing shrinkage	%	≤0.4
Compressive strength of the resin after curing	MPa	≥50
(50% deformation)	MPa	≥10
Shear strength	MPa	≥10

Step 5: Epoxy resin curing process

The epoxy resin needs to be cured for a week, and then the rubber sealing tapes can be removed, as shown in Figure 10.

Figure 10. SSESS after curing for one week.

2.4. Hydraulic Burst Test

The hydraulic burst test is commonly used in the design verification of pipes and pipe fittings [20]. To verify the repair reliability of the SSESS, the two hydraulic burst tests for two tees were performed. One of the two tees was not repaired, and the other was repaired with the SSESS, as shown in Figure 11. The strain gauges were arranged on both tees according to the same layout of strain gauges on the outer surface of the tees. In addition, these conditions, such as material properties, welding parameters, and geometric parameters of structural parts of the tee hydrostatic test were consistent with each other, except for the repairing approaches. The two hydraulic burst tests were conducted at the HYDROSEYS system.

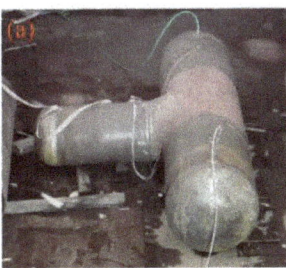 

Figure 11. Hydraulic test structural part: (**a**) unrepaired tee, and (**b**) SSESS repaired tee.

2.5. Simulations

To analyze the stress and strain distributions of the tee by the repairment of the special-shaped epoxy resin steel sleeve, three-dimensional FE models of the hydraulic test structural parts were carried out by ABAQUS/Explicit software. The load condition of the two simulations only considered the effect of hydrostatic pressure from 0 to burst pressure. The three ends and the outer surface were free, which is consistent with the hydraulic test.

The lengths of three straight pipes all were 1100 mm. The straight pipes were meshed using three-dimensional reduced-integration 8-node solid elements (C3D8R) and the tees were meshed using three-dimensional 10-node modified quadratic tetrahedron elements (C3D10M). The structural parts of straight pipes and the tee were created in an overall model instead of a combination of these parts. The three straight parts and the tee part were meshed by C3D8R and C3D10M element type, respectively. Then, the elements of C3D8R and C3D10M were bonded by operating the "mesh part" function automatically in the ABAQUS software, as shown in Figure 12. The yield strengths of the two tees were measured by the indentation, metallurgic replica, and Leeb hardness technologies. The yield strength of the tee is ~350 MPa, which is lower than the 485 MPa requirement. The material constitutive model used the elasto-plastic hardening material models of S1, S2, and S3 in Section 2.1 by fitting the stress–strain data. The special form of these material models can be expressed as:

$$\sigma = \begin{cases} E\varepsilon_e, for\ \sigma \leq \sigma_s \\ K\varepsilon_p^n, for\ \sigma \geq \sigma_s, \end{cases} \quad (2)$$

where E is Young's modulus, σ_s is the yield stress, ε_e is the elastic strain, ε_p is the plastic strain, K is the strength coefficient, and n is the strain-hardening exponent. The constitutive parameters of S1, S2, and S3 parts are listed in Table 3.

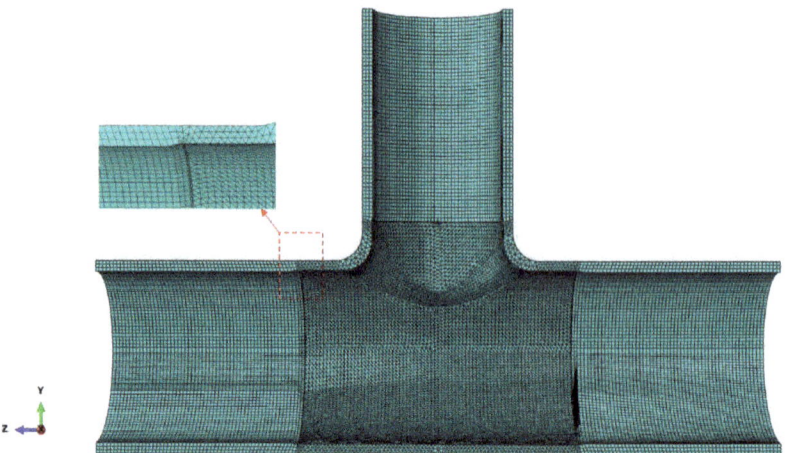

Figure 12. Mesh design of the tee and straight parts.

Table 3. Properties of S1, S2, and S3 parts.

Region of Tee	E (GPa)	K (MPa)	n
S1	207	281.0	0.295
S2	211	183.5	0.295
S3	209	271.0	0.275

3. Results

Figure 13 shows the simulation of Mises stress contours with different L_h at P = 12 MPa. As shown in Figure 13, the regions of the tee end and girth weld with stress concentrations as L_h are -175 mm and -100 mm, while the stress concentration of the tee end gradually reduces with the increase in L_h. Therefore, the design of L_h for the special-shaped steel sleeve should be larger than 100 mm.

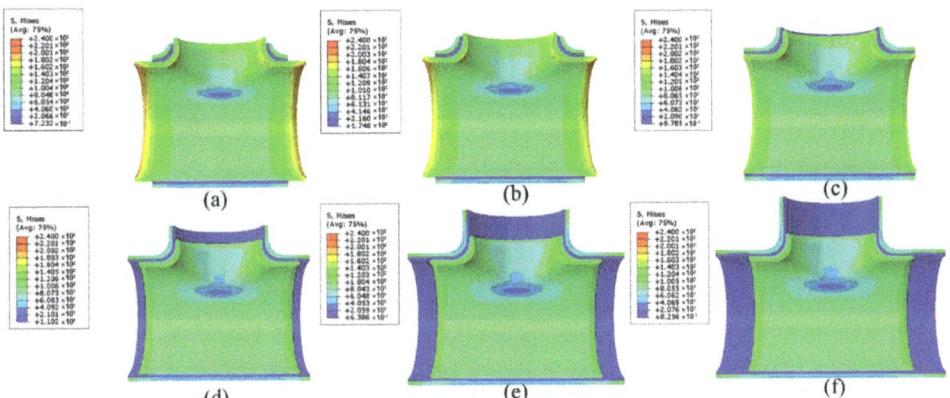

(a) (b) (c)
(d) (e) (f)

Figure 13. Simulation Mises stress contours with different L_h at P = 12 MPa: (**a**) −175 mm; (**b**) −100 mm; (**c**) 0 mm; (**d**) 100 mm; (**e**) 200 mm; (**f**) 300 mm.

Figure 14 shows the comparisons of strain at monitoring locations between FEM and the experiment for unrepaired and repaired tees. As shown in Figure 14, the strains at monitoring locations are consistent with the experimental strain results for unrepaired and repaired tees, which verified the correctness and reliability of unrepair and repair finite element models.

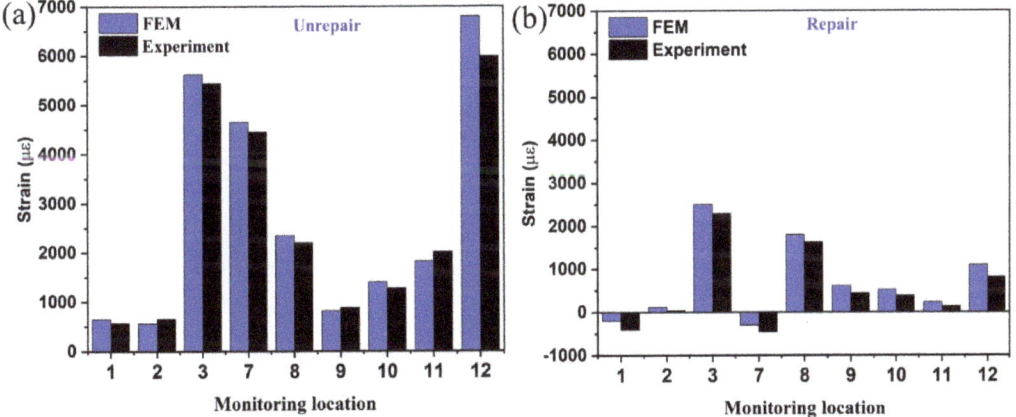

Figure 14. Comparisons of strain at monitoring locations between FEM and the experiment: (**a**) unrepaired and (**b**) repaired tees.

The equivalent stress contour of unrepaired and repaired tees for inner and outer surfaces under a hydrostatic pressure of 24 MPa are exhibited in Figures 15 and 16, respectively. Figure 17 shows the Mises stress tendencies of critical regions (shoulder and belly of tee) between unrepaired and repaired tees with the increase in internal pressure for inner and outer surfaces. As shown in Figures 15–17, the stress concentration was mainly distributed on the inner surface of the shoulder and the outer surface of the belly of the tee, while the stress concentration was greatly alleviated by the repair of an SSESS repair. The maximum stress of the inner shoulder surface and outer belly surface exceeded the yield stress of the tee, so the stress concentration is dangerous for the safe operation of the tee. The reduced stress concentration areas decrease the possibility of failure for the low-strength tee.

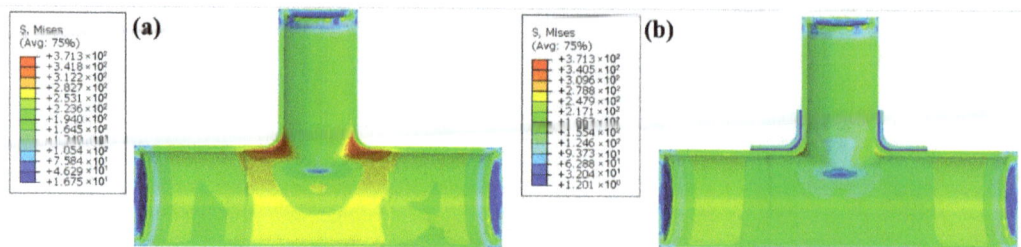

Figure 15. Stress contour of the inner surface of low strength under a hydrostatic pressure of 24 MPa: (**a**) unrepair, and (**b**) SSESS repair.

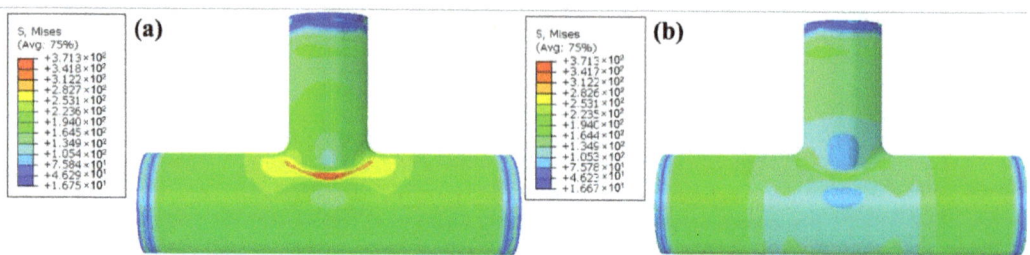

Figure 16. Stress contour of the outer surface of low strength under a hydrostatic pressure of 24 MPa: (**a**) unrepair, and (**b**) SSESS repair.

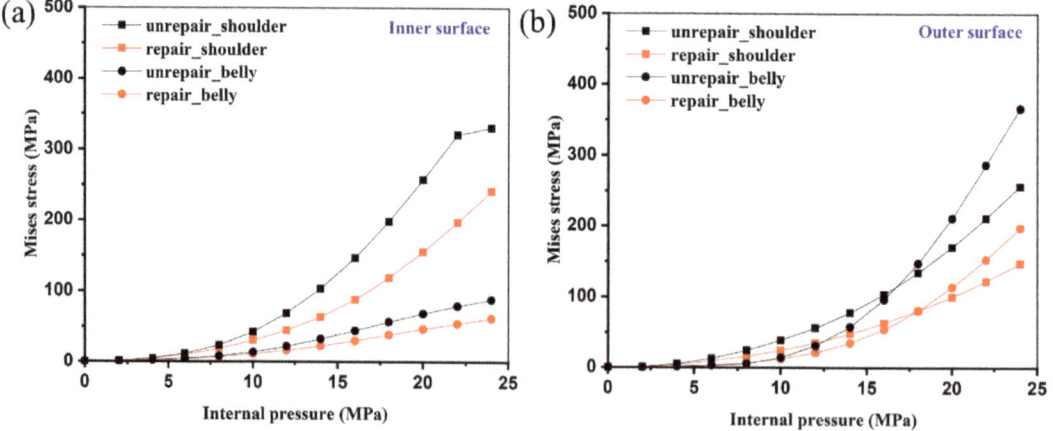

Figure 17. Mises stress tendencies of critical regions (shoulder and belly of tee) between unrepaired and repaired tees with the increase in internal pressure: (**a**) inner surface and (**b**) outer surface.

Figure 18 shows the experimental hydraulic burst curves for the unrepaired and repaired tees. As shown in Figure 18, the yielding point could be distinguished from the pressure–time curves. The hydraulic and burst pressures of the two cases are ~32 MPa, ~48.6 MPa, and ~42 MPa, ~50.8 MPa, respectively. It can be judged from the pressure-time curve of the repair tee and the failure morphology of the sleeve that the special-shaped epoxy resin steel sleeve did not work under 42 MPa of hydrostatic pressure, as shown in Figures 18 and 19.

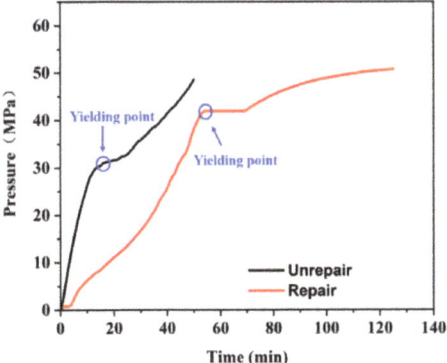

Figure 18. Experimental hydraulic burst curves of unrepaired and SSESS repaired.

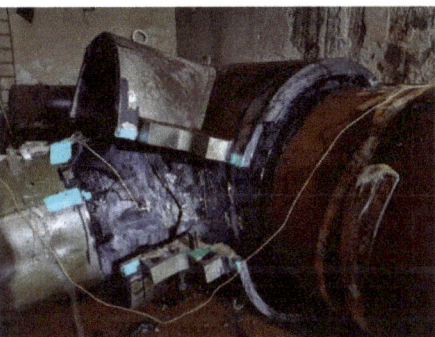

Figure 19. Failure of SSESS with bolt fracture.

As shown in the simulations results in Figures 20 and 21, the maximum equivalent stress areas of the unrepaired and repaired tees are both in the belly, signifying that the origin of the burst should be in the belly. As the experimental burst appears in the two tees in Figures 20 and 21, the origin of the burst of unrepaired and repaired tees are both in the belly and cracked and expanded in the X direction, which proved the reliability of FE models. The shoulder and the girth weld of the branch tee are both torn during the cracking process.

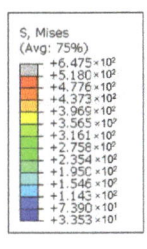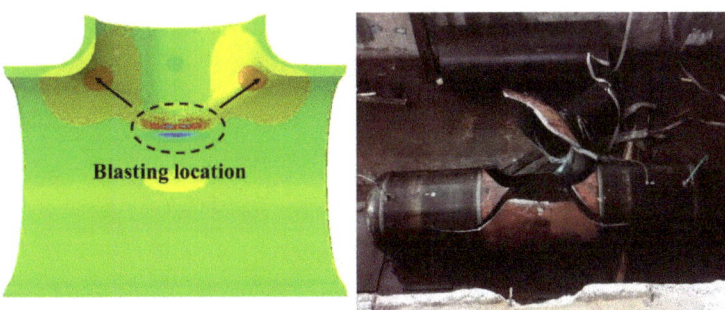

Figure 20. Simulated prediction and experimental verification of the blasting location for the unrepaired low-strength tee.

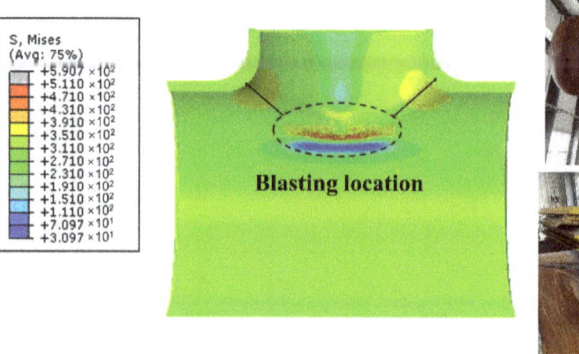
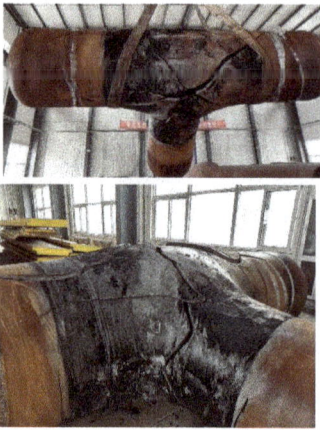

Figure 21. Simulated prediction and experimental verification of the blasting location for the SSESS repaired low-strength tee.

4. Discussion

The experimental strain monitoring results of unrepair and repair conditions under the hydrostatic pressure of 18 MPa and 24 MPa show that the strain concentrations are mainly distributed in the shoulder (location 3), belly (location 12), and the welding joint of the branch pipe (location 7 and location 8), as shown in Figure 22. The strain concentration locations follow the simulation results. The repairing result of the special-shaped epoxy resin steel sleeve shows that the strain concentrations were greatly reduced, especially for the belly (location 12). The maximum strain is ~0.006 on the outer surface of the belly, and it decreases from ~0.006 to ~0.001 under 24 MPa of hydrostatic pressure; the decreasing rate is about 80%. The hydrostatic pressures of 18 MPa and 24 MPa represent $1.5 \times P_d$ (1.5 times the value of the design pressure P_d) and $2 \times P_d$ (twice the value of the design pressure P_d), respectively; $1.5 \times P_d$ is the pressure during the test run, and $2 \times P_d$ represents a design coefficient of 0.5, which is a requirement at the natural gas station. Therefore, the stress and strain measurements under 18 MPa and 24 MPa are conducive to analyzing the load capacity under the test run and the required design coefficient.

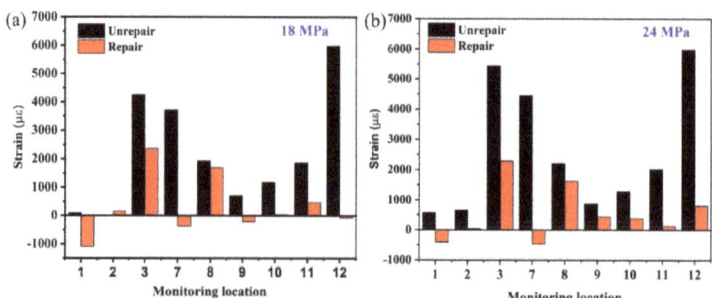

Figure 22. Comparisons of strain monitoring between unrepaired and SSESS repaired tees: (a) 18 MPa, and (b) 24 MPa.

The yielding pressure is mainly associated with the yielding point of the loading curves. It is noted that the yielding and burst pressures were improved from ~32 MPa and ~48.6 MPa to ~42 MPa and ~50.8 MPa, respectively, which demonstrates the reliability of a special-shaped epoxy resin steel sleeve to repair a low-strength tee. The reason for the burst pressure of the unrepaired tee being greater than the repair tee is the difference in the material properties and geometry. Moreover, the special-shaped epoxy resin steel sleeve did not work before the burst pressure.

For the low strength tee, the weak points of bearing deformation mainly exist in the shoulder, belly, and girth weld of the tee. If the welding quality of the girth weld is excellent, the reinforcement of the tee should focus on eliminating the stress concentration effect on the shoulder and abdomen. The main functions of the profiled epoxy steel sleeve are to (1) reduce the stress concentration effect of the low strength tee and (2) improve the safety factor and deformation resistance of the tee bearing. The stage in which the special-shaped epoxy steel sleeve plays a role should be the elastic stage of the whole reinforcing structure. Once it enters the elastic-plastic deformation stage, the epoxy resin glue and bolts between the special-shaped epoxy steel sleeves will fail.

5. Conclusions

To repair the low-strength tee, an SSESS repair method was proposed. According to the design of the special-shaped steel sleeve, the repairing test, the hydraulic burst test, and simulations, the repairing reliability of the special-shaped epoxy resin steel sleeve was proved. The main conclusions drawn are as follows:

(1) The critical design parameters for the special-shaped sleeve are the minimum thickness, the material properties of the sleeve, the length between the edge of the sleeve and the girth weld, the gap between the sleeve and the tee, and the locations of the injection and exhaust holes.
(2) The strain concentrations of the unrepaired tee were greatly reduced, especially for the belly. The maximum strain was ~0.006 on the outer surface of the belly, and it decreased from ~0.006 to ~0.001 under hydrostatic pressure of 24 MPa, where the decreasing rate was about 80%.
(3) The yielding and burst pressure was improved from ~32 MPa and ~48.6 MPa to ~42 MPa and ~50.8 MPa, respectively, which demonstrates the good reliability of the special-shaped epoxy resin steel sleeve repairing low-strength tee.

Author Contributions: Conceptualization, J.C., W.M. and H.J.; methodology, J.C.; software, J.C. and H.N.; validation, T.Y. and J.R.; formal analysis, H.N., X.L. and W.D.; investigation, J.C.; data curation, J.C. and X.L.; writing—original draft preparation, J.C.; writing—review and editing, K.W. and H.J.; supervision, W.M.; project administration, H.J.; funding acquisition, J.C. All authors have read and agreed to the published version of the manuscript.

Funding: This research was funded by the Young Scientists Fund of the National Natural Science Foundation of China [grant number 51904332], the Natural Science Foundation of Shaanxi Province, China [grant number 2020JQ-934], and the Scientific research and technology development projects of CNPC [grant number 2019D-5009-13].

Data Availability Statement: Not applicable.

Conflicts of Interest: The authors declare no conflict of interest.

References

1. Cao, J.; Ma, W.; Pang, G.; Wang, K.; Yao, T. Failure analysis on girth weld cracking of underground tee pipe. *Int. J. Press. Vessel. Pip.* **2021**, *191*, 104371. [CrossRef]
2. Su, W.; Luo, W.; Dong, X.; Liu, R. Ultimate Strength Analysis of Local Thinning Tee Pipe Considering Plastic Strengthening Effect. *J. Phys. Conf. Ser.* **2020**, *1650*, 022024. [CrossRef]
3. Shuai, Y.; Wang, X.; Wang, J.; Yin, H.G.; Cheng, Y.F. Modeling of mechanical behavior of corroded X80 steel pipeline reinforced with type-B repair sleeve. *Thin Walled Struct.* **2021**, *163*, 107708. [CrossRef]
4. Jaske, C.E.; Hart, B.O.; Bruce, W.A. *Pipeline Repair Manual*; Pipeline Research Council International: Arlington, VA, USA, 2006; pp. 1–196.
5. American Society of Mechanical Engineers. *Repair of Pressure Equipment and Piping*; American Society of Mechanical Engineers: New York, NY, USA, 2015.
6. Lim, K.S.; Azraai, S.; Yahaya, N.; Noor, N.M.; Kim, J. Behaviour of steel pipelines with composite repairs analysed using experimental and numerical approaches. *Thin Walled Struct.* **2019**, *139*, 321–333. [CrossRef]
7. Rinastiti, M. *The Effect of Aging on Composite-to-Composite Repair Strength*; Citeseer: State College, PA, USA, 2010.

8. Savari, A.; Rashed, G.; Eskandari, H. Time-Dependent Reliability Analysis of Composite Repaired Pipes Subjected to Multiple Failure Modes. *J. Fail. Anal. Prev.* **2021**, *21*, 2234–2246. [CrossRef]
9. Karpenko, M.; Prentkovskis, O.; Šukevičius, Š. Research on high-pressure hose with repairing fitting and influence on energy parameter of the hydraulic drive. *Eksploat. Niezawodn. Maint. Reliab.* **2021**, *24*, 25–32. [CrossRef]
10. Batisse, R. Review of gas transmission pipeline repair methods. In *Safety, Reliability and Risks Associated with Water, Oil and Gas Pipelines*; Springer: Berlin/Heidelberg, Germany, 2008; pp. 335–349.
11. Stephens, D.; Leis, B.; Francini, R. Developments in criteria for repair or replacement decisions for pipeline corrosion defects. In Proceedings of the 3 Congreso y Expo Internacional de Ductos, Monterrey, Mexico, 13–15 December 1998; pp. 268–275.
12. Wood, A. Evolution of epoxy sleeve pipeline repair technology. *Pipes Pipelines Int.* **2015**, *23*, 50–52.
13. *CSA-Z662-2007*; Oil and Gas Pipeline Systems. Canadian Standards Association: Toronto, ON, Canada, 2007.
14. Novák, P.; Žmindák, M.; Pelagić, Z. High-pressure pipelines repaired by steel sleeve and epoxy composition. In *Applied Mechanics and Materials*; Trans Tech Publications Ltd.: Bäch, Switzerland, 2014; pp. 181–188.
15. Mazurkiewicz, L.; Tomaszewski, M.; Malachowski, J.; Sybilski, K.; Chebakov, M.; Witek, M.; Yukhymets, P.; Dmitrienko, R. Experimental and numerical study of steel pipe with part-wall defect reinforced with fibre glass sleeve. *Int. J. Press. Vessel. Pip.* **2017**, *149*, 108–119. [CrossRef]
16. Arif, A.; Al-Nassar, Y.N.; Al-Qahtani, H.; Khan, S.; Anis, M.; Eleiche, A.; Inam, M.; Al-Nasri, N.I.; Al-Muslim, H.M. Optimization of pipe repair sleeve design. *J. Press. Vessel. Technol.* **2012**, *134*, 51702. [CrossRef]
17. Jaszak, P.; Skrzypacz, J.; Borawski, A.; Grzejda, R. Methodology of Leakage Prediction in Gasketed Flange Joints at Pipeline Deformations. *Materials* **2022**, *15*, 4354. [CrossRef] [PubMed]
18. Shang-Peng, L.; Lei, Y.; Xuan-Peng, Q.; Xuefei, D.; Weihua, X.; Huatian, X.; Zhenhua, C. The Applicability Study of Special Steel Sleeve Filled with Epoxy Resin for the Leakage of Welding Connection Platform on Brine Pipeline. *Total Corros. Control* **2016**, *8*, 57–59.
19. Ahmad, K.; Baig, Y.; Rahman, H.; Hasham, H. Progressive failure analysis of helicopter rotor blade under aeroelastic loading. *Aviation* **2020**, *24*, 33–41. [CrossRef]
20. Fowler, J.R.; Alexander, C.R. *Design Guidelines for High-Strength Pipe Fittings*; American Gas Association, Inc.: Arlington, VA, USA; Stress Engineering Services: Houston, TX, USA, 1994.

Article

The Effect of Strain Rate on the Deformation Behavior of Fe-30Mn-8Al-1.0C Austenitic Low-Density Steel

Jiahui Du [1], Peng Chen [1,2,*], Xianjun Guan [1], Jiawei Cai [1], Qian Peng [1], Chuang Lin [1] and Xiaowu Li [1,3,*]

1. Department of Materials Physics and Chemistry, School of Materials Science and Engineering, Northeastern University, Shenyang 110819, China
2. Key Laboratory for Anisotropy and Texture of Materials, Northeastern University, Shenyang 110819, China
3. State Key Laboratory of Rolling and Automation, Northeastern University, Shenyang 110819, China
* Correspondence: chenpeng@mail.neu.edu.cn (P.C.); xwli@mail.neu.edu.cn (X.L.)

Abstract: Automotive steels suffer different strain rates during their processing and service. In this study, the effect of strain rates on the tensile properties of fully austenitic Fe-30Mn-8Al-1.0C (wt.%) steel was investigated, and the dominant deformation mechanism was clarified. Conventional and interrupted tension tests and various microscopic characterization methods were carried out in this study. The results indicate that the yield strength increases with the increasing strain rate in the range of 10^{-4}–10^{-1} s^{-1}, and a good strength–ductility combination was achieved in the sample deformed at 10^{-3} s^{-1}. In the process of straining at 10^{-3} s^{-1}, microbands and deformation twins were observed. Thus, the combination of microband induced plasticity (MBIP) together with twinning induced plasticity (TWIP) leads to a continuous strain hardening behavior, and consequently to superior mechanical properties. However, adiabatic heating that leads to the increase in stacking fault energy (SFE) and inhibits the TWIP effect, as well as thermal softening jointly induces an anomalous decrease in tensile strength at the high strain rate of 10^{-1} s^{-1}.

Keywords: strain rate; mechanical property; austenitic low-density steel; strain hardening; microstructure; deformation mechanism

---

Citation: Du, J.; Chen, P.; Guan, X.; Cai, J.; Peng, Q.; Lin, C.; Li, X. The Effect of Strain Rate on the Deformation Behavior of Fe-30Mn-8Al-1.0C Austenitic Low-Density Steel. *Metals* 2022, 12, 1374. https://doi.org/10.3390/met12081374

Academic Editor: Ruslan R. Balokhonov

Received: 21 July 2022
Accepted: 15 August 2022
Published: 18 August 2022

Publisher's Note: MDPI stays neutral with regard to jurisdictional claims in published maps and institutional affiliations.

Copyright: © 2022 by the authors. Licensee MDPI, Basel, Switzerland. This article is an open access article distributed under the terms and conditions of the Creative Commons Attribution (CC BY) license (https://creativecommons.org/licenses/by/4.0/).

1. Introduction

As one kind of potential structural steel in the automotive industry, Fe-Mn-Al-C low-density steel has attracted increasing scientific and commercial attention, because of its light-weight which satisfies the requirements in increasing the fuel efficiency and reducing the gas emissions of automobiles [1–5]. Al alloying addition in the Fe-Mn-Al-C steels has an effect on density reduction, and per 1 wt.% Al reduces the density by 1.3% [4]. The Fe-Mn-Al-C steels could be either ferrite, austenite or ferrite-austenite duplex, depending mainly on the relative content of alloying elements [5]. Fully austenitic Fe-Mn-Al-C steels, with a high Al content in the range from 5 to 12 wt.%, Mn content between 12 and 30 wt.%, and C content between 0.6 and 2.0 wt.%, exhibit extraordinary strength-ductility combinations (UTS: 0.6–1.5 GPa; elongation: 30–100%), and the austenite phase is very stable during deformation [6–9].

During the production or service process, automotive steels are normally subjected to various strain rates. For instance, automobile steels suffer the strain rate from 10^{-1} to 10 s^{-1} during the forming process and 10^2 to 10^3 s^{-1} in the event of a collision. Yoo et al. [10] investigated the strain rate sensitivity of austenitic Fe-28Mn-9Al-0.8C steels. It was found that the yield strength increased with an increasing strain rate from 2×10^{-4} to 10^{-1} s^{-1}, exhibiting an overall positive strain rate sensitivity. In contrast, the ductility decreased with an increasing strain rate, due to the thermal activation in this strain rate range. Furthermore, the sensitivity of the strain rate for Fe-22Mn-0.6C-1.5Al twinning-induced plasticity steel has been also investigated by Yang et al. [11], who found that the ultimate tensile strength and the uniform elongation decreased with increasing strain rate from 10^{-4} to 1 s^{-1}; it is related to the fact that the deformation twins at a lower strain rate were much thinner and denser than those at a higher strain rate. Although there are some investigations on the relationship between the mechanical properties and strain rate of austenitic Fe-Mn-Al-C steels, the deformation mechanisms regarding microstructure evolution still remains unclear.

Accordingly, the present study aims to elucidate the effect of strain rate on the tensile deformation behavior of fully austenitic Fe-30Mn-8Al-1.0C (wt.%) steel. The tensile tests of solid solution samples were conducted at various strain rates. Detailed microstructural observations were performed on the lateral surface and fracture surface of the failed samples by scanning electron microscopy (SEM). For clarifying the evolution of dislocation substructures and their interactions, the microstructure of the deformed samples at the strain rate of 10^{-3} s^{-1} was characterized by transmission electron microscopy (TEM).

2. Materials and Methods

The ingots with the compositions of Fe-30Mn-8Al-1.0C (wt.%) were prepared by induction melting in a vacuum smelting furnace. In order to remove segregation zones originating from solidification, the ingots were homogenized at 1200 °C for 3 h and hot-rolled to the thickness of 5 mm by multiple passes at a final temperature exceeding 850 °C, and then cooled to room temperature. Afterwards, the hot-rolled sheets were solution treated at 1050 °C for 1 h, followed by water quenching. The density was measured to be 6.8 g·cm^{-3} according to the Archimedes principle.

Tensile samples with a gauge dimension of 13 mm × 5 mm × 2 mm were machined along the rolling direction. The tensile samples were initially polished using #2000 SiC paper and then electrochemically polished using 10% perchloric acid alcohol with a voltage of 25 V for 25 s. Tensile tests were performed on the samples up to failure using a universal testing machine AG-Xplus250kN (SHIMADZU, Kyoto, Japan) with the strain rates of 10^{-4}, 10^{-3}, 10^{-2} and 10^{-1} s^{-1} at room temperature. Except tensioning to fracture, interrupted tensile tests up to predetermined true strains ε_T of 1%, 10% and 25% were carried out at the strain rate of 10^{-3} s^{-1} for tracking the microstructural evolution during tensile deformation. For ensuring the accuracy of tensile properties, at least three tests were repeated for each condition.

The initial solution-treated samples were polished and etched in a solution of 100 mL methanol, 1 mL hydrochloric acid and 4 g picric acid, and then observed by an optical microscope Axio Imager A1m (ZEISS, Oberkochen, Germany). To reveal the microscopic deformation mechanism, the uniform deformation zone and fracture morphology of failed samples were observed using a scanning electron microscope JSM-6510A (JEOL, Tokyo, Japan). After interrupted deformation, the slices with 600 μm thick were spark cut from the deformed part of tensile samples and manually polished to the thickness of 50 μm. Then, the TEM slices were perforated under an electrolytic double spray thin-reducing instrument with 6% perchloric acid alcohol at 20 V and −30 °C. The microstructure characterization was performed by a transmission electron microscope Tecnai G20 (FEI, Hillsbro, OR, USA).

3. Results and Discussion

3.1. Initial Solution-Treated Microstructure

Figure 1 shows a representative optical micrograph of Fe-30Mn-8Al-1.0C steel. As illustrated in Figure 1a, uneven equiaxed fine grains and some annealing twins were observed in hot-rolled steel. In order to eliminate the residual stress in hot-rolled steel and obtain uniform single-phase austenite structure, the samples were solution treated at 1050 °C for 1 h, and the average grain size of the solid solution treated steel is calculated to be approximate 35 μm (Figure 1b). Single austenite phase with a great deal of annealing twins (up to 57%) could be observed. C and Mn are austenite stabilizing elements, and the addition of high content of Mn and C results in a stable single austenite phase. In addition, Al has a suppressive effect on the precipitation of some carbides (such as cementite), inducing more C enrichment in austenite [12]. The annealing twins could effectively improve the properties of metallic materials by increasing the proportion of austenite grain boundary and refining the austenite grains.

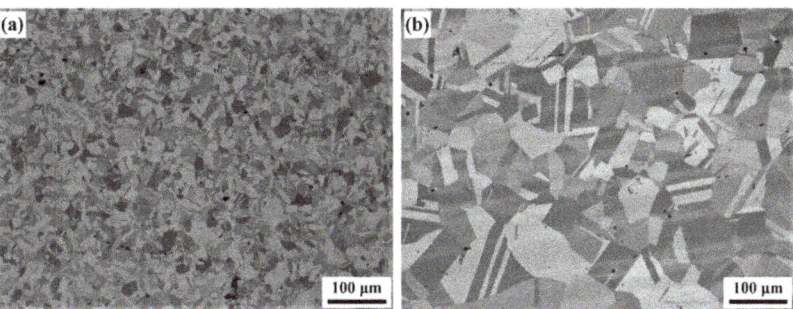

Figure 1. Optical micrograph of Fe-30Mn-8Al-1.0C steel. (**a**) Hot-rolled steel; (**b**) solid solution steel treated at 1050 °C for 1 h.

3.2. Tensile Properties

In order to examine the effect of the strain rate on the mechanical properties, uniaxial tensile tests were conducted on the solution-treated Fe-30Mn-8Al-1.0C steel at strain rates ranging between 10^{-4} and 10^{-1} s^{-1}. The representative stress–strain curves and the strain hardening rate curves of the steels tested at room temperature with different strain rates are presented in Figure 2. The mechanical properties are listed in Table 1. The studied steel exhibits a continuous yield behavior and a high strain hardening ability during tensile deformation under all strain-rate conditions. The yield strength increases with increasing strain rate, which are 301 ± 30, 329 ± 20, 378 ± 5 and 381 ± 8 MPa at the strain rates of 10^{-4}, 10^{-3}, 10^{-2} and 10^{-1} s^{-1}, respectively. The ultimate tensile strength exhibits an uptrend until 10^{-2} s^{-1} followed by a decline at 10^{-1} s^{-1}, showing a maximum of 801 ± 12 MPa. The total elongation is as high as 72 ± 3% at the strain rate of 10^{-3} s^{-1}. Generally, the ductility becomes weak with enhancing strength for metal materials, and the product of ultimate tensile strength and total elongation (UTS × e_f) is a comprehensive performance index to characterize the strength and ductility of metallic materials. The sample deformed at the strain rate of 10^{-3} s^{-1} achieves the highest product of strength and ductility of 56,800 MPa%.

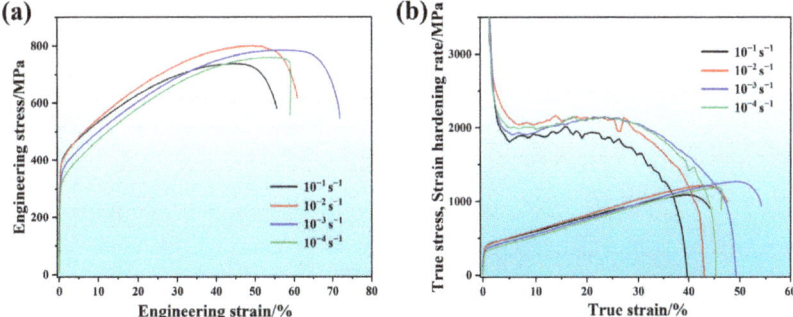

Figure 2. (**a**) The engineering stress–strain curves of the solution-treated Fe-30Mn-8Al-1.0C steel tested at room temperature with different strain rates, and (**b**) the true stress–strain curves and work hardening rate curves of the studied steel.

Table 1. Mechanical properties of the solution-treated Fe-30Mn-8Al-1.0C steel at different strain rates.

Strain Rate/s^{-1}	Yield Strength/MPa	Ultimate Tensile Strength/MPa	Total Elongation/%
10^{-4}	301 ± 30	761 ± 35	59 ± 2
10^{-3}	329 ± 20	789 ± 25	72 ± 3
10^{-2}	378 ± 5	801 ± 12	61 ± 1
10^{-1}	381 ± 8	738 ± 9	56 ± 3

Lower ultimate tensile strength and total elongation are exhibited when the sample deformed at the strain rate of 10^{-1} s^{-1}, which is probably related to the thermal softening effect at high strain rates. Quasi-static tensile tests at the strain rate range of 10^{-5} to 10^{-2} s^{-1} are generally regarded as isothermal process, and it is difficult to dissipate heat to the surrounding above this strain rate range [13,14]. Thus, the adiabatic heating effect should be considered while the sample is deformed at higher strain rates. The adiabatic temperature rise ΔT can be defined by [15]:

$$\Delta T = \frac{\Delta Q}{\rho C_p} = \frac{\beta}{\rho C_p} \int_{\varepsilon_1}^{\varepsilon_2} \sigma d\varepsilon \quad (1)$$

where ΔQ is the energy converted from mechanical energy to thermal energy, which is obtained by integrating the true stress-strain curve, β is the coefficient of thermal energy converted from mechanical energy, ρ is the density of the present steel and C_p is the typical specific heat capacity. According to Equation (1), the temperature of the sample deformed at the strain rate of 10^{-1} s^{-1} increases by about 101 °C. For FCC materials with the high-concentration solid solution element, forest dislocations and short-range order (SRO) are the main obstacles for dislocation glide [16,17]. It is logical that the increase in the strain rate could enhance the dislocation mobility, and the dislocations would overcome obstacles more easily with the assistance of thermal energy. Additionally, the dynamic recovery of dislocations is accelerated at an increased temperature, which causes a negative effect on the strain hardening. Thereby, the strain hardening capability is weakened in the steel deformed at high strain rate. As presented in Figure 2b, the sample deformed at 10^{-1} s^{-1} has the lowest strain hardening rate, inducing the lowest strength and ductility. In addition, the sample deformed at 10^{-3} s^{-1} shows the highest strain hardening ability, and consequently the best combination of strength and ductility.

3.3. Morphologies of Deformation and Fracture

The fracture surfaces of the samples deformed at different strain rates were observed by SEM, as illustrated in Figure 3. The tensile fracture characteristics of the studied steels are similar, all involving fibrous zone, radial zone and shear lip zone (Figure 3a–d). The fibrous zone accounts for the largest proportion of the fracture surface. All of the fracture surface are composed of equiaxed dimples with a bimodal size formed by a microvoid coalescence, exhibiting the characteristic of ductile fracture mode (Figure 3e–h). Thus, the samples deformed at four strain rates all exhibit good ductility (Figure 2).

In addition, the distinct "snake slips" or "ripples" were observed in the inner walls of some large-sized dimples, indicating that severe plastic deformation and localized strain induce new dislocations to slip on the surface of the dimples [18]. By comparison, it is found that diameter and depth of the dimples reach a maximum in the sample deformed at 10^{-3} s^{-1}. These features indicate that the ductility of the sample deformed at 10^{-3} s^{-1} is better, which is consistent with the experimental results of mechanical properties.

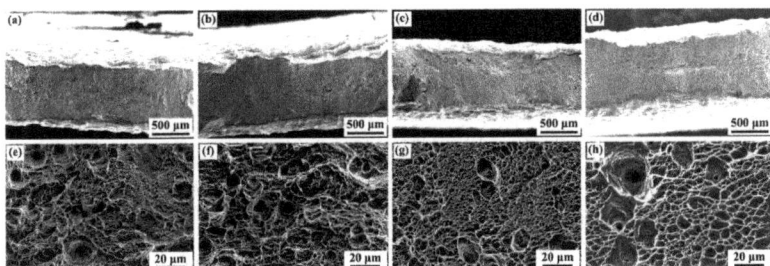

Figure 3. SEM fractographs of the samples tensioned at different strain rates of 10^{-4} s^{-1} (**a,e**), 10^{-3} s^{-1} (**b,f**), 10^{-2} s^{-1} (**c,g**), and 10^{-1} s^{-1} (**d,h**).

Figure 4 illustrates the SEM micrographs of the lateral surface in the uniform deformation region and near fracture. At uniform deformation region, the austenite grains are elongated, and the dislocation slip characteristics are observed under all strain rates (Figure 4a–d). The dislocation slipping and the grain rotation contribute to coordinate the plastic deformation [19]. All observations reveal that there are some extrusions caused by dislocation slipping, and some microvoids are generated by the slipping extrusion at grain boundaries. Moreover, a part of slip bands pass through annealing twins to form "Z" shape slipping. For FCC materials, the stacking of {111} planes in the disrupted sequence is called a stacking fault, which is a two-dimensional defect formed by the nucleation and propagation of partial dislocations. The stacking fault has an associated energy called stacking fault energy (SFE, with units of J·m^{-2}), since the atomic packing within it is no longer typical of the FCC structure. At the high strain rate larger than 10^{-3} s^{-1}, the adiabatic heating during deformation might increase the SFE, which has a suppressive effect on the planar slip of dislocations. On the sites near the fracture, the lateral surface becomes rougher, and the extrusion induced by dislocation slipping becomes more and more apparent (Figure 4e–f), indicating more severe plastic deformation.

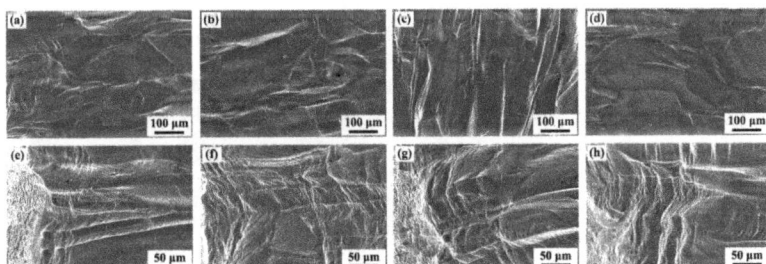

Figure 4. SEM images exhibiting the surface features at uniform deformation zones (**a–d**) and corresponding surface features near the fracture surfaces (**e–h**) of tensile samples at the strain rates of 10^{-4} s^{-1} (**a,e**), 10^{-3} s^{-1} (**b,f**), 10^{-2} s^{-1} (**c,g**) and 10^{-1} s^{-1} (**d,h**).

A few previous studies [20–22] suggested that inter/intra-granular κ-carbides and ordered L1$_2$ phase can be formed in solution-treated Fe-Mn-Al-C steels despite the absence of aging. Considering the promoting effect of Al on the precipitation of κ-carbides and high Al addition, κ-carbides should exist and affect the mechanical properties of the present steel [23,24]. A coherency strain field around intragranular κ-carbides have a considerable influence on impeding the dislocation motion, resulting in the strengthening effect. By contrast, the existence of intergranular κ-carbides is conducive to the initiation and propagation of cracks, which contributes to the fracture [25]. Furthermore, L1$_2$ phase is regarded as short-range order (SRO), which is found to promote the planar glide of dislocations. It is usually considered that SRO could be sheared and destroyed by the leading dislocations, and then the subsequent dislocations can slip across the destroyed

SRO region more easily. This effect causes the planar glide sufficiently, which is named as "glide plane softening" [17,26–29].

3.4. Microstructure Evolution during Deformation

In order to understand the microstructural evolution and related deformation mechanism during the tensile process, the interrupted tensile tests and TEM observations were conducted. Considering its good comprehensive performance of strength and ductility, the samples deformed at 10^{-3} s^{-1}, up to ε_T = 1%, 10%, 25% and failure were selected for the examinations, respectively. The strain hardening rate as a function of true strain exhibits an obvious three-stage strain hardening behavior, including quick dropping, recovery and slow decreasing of strain hardening rate, as shown in Figure 5a.

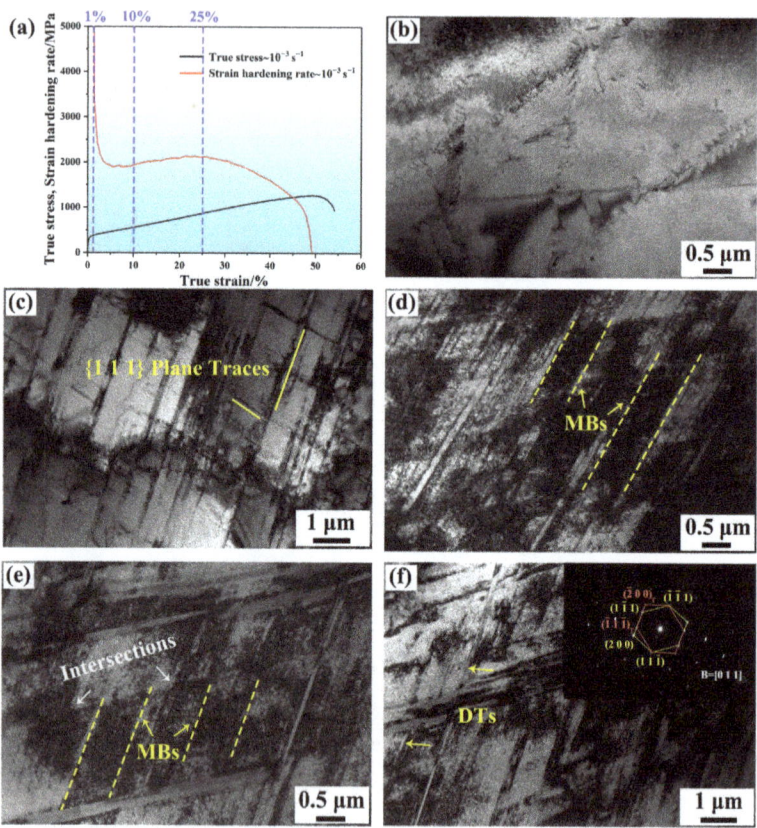

Figure 5. True stress-strain curves and strain hardening rate curves of the sample deformed at the strain rate of 10^{-3} s^{-1} (**a**), together with TEM micrographs of the present steel deformed to the true strains of 1% (**b**), 10% (**c**), 25% (**d**) and 100% (or final fracture) (**e**,**f**).

Figure 5b shows early-formed dislocation arrays aligned along the slip direction at the early deformation stage of 1% strain, which is a typical kind of planar slip dislocation configuration in FCC metals. The pronounced planar slip behavior of dislocations results in the formation of crystallographically aligned slip bands. After straining to 10%, the microstructure is still characterized by planar slip bands with a slip band spacing of approximate 700 nm (Figure 5c). The observation indicates the slip trace of two non-coplanar {111} slip planes, forming a Taylor-lattice structure, which is a kind of low energy dislocation structure [30,31]. Therefore, the planar slip of dislocations dominates the deformation mechanism at low strains.

After straining to 25%, extensive parallel high-density slip bands and microbands are found in the matrix, and the spacing of slip bands is less than 100 nm (Figure 5d). The slip bands have been evolved into high-density dislocation walls (HDDWs), which are composed of the dense dislocation bands lying on or parallel to the most active slip systems. Actually, in FCC alloys, Shockley partial dislocations slip on {111} planes and they attract or repulse each other, which are determined by the Burgers vector directions of the leading partials. If one of them is trapped by a HDDW and attracts one another, they can fuse into a sessile dislocation. During this process, further dislocation slip is hindered on both {111} planes. This causes the increase in density of HDDWs, which contributes to the barrier to the dislocation motion [32]. Moreover, Taylor lattices rotate to accommodate further strain, and a domain boundary develops from an original HDDW as a result of the misorientation between Taylor lattice domains. In the same way, a second domain boundary develops parallel to the previous one, and a microband manifested by misorientations on both sides takes shape [1,33–35].

As the strain amount further increases, well-developed microbands and their intersections are observed in fractured sample (Figure 5e). Microbands are characterized by dense tangled dislocations with a lower activity and they are usually aligned along the trace of {111} slip planes in no cell-forming FCC alloys. The microbands and their intersections can subdivide the grains. The high-density dislocations in the microbands hinders dislocation motion and decrease the effective free path of movable dislocations, resulting in the Hall-Petch strengthening effect [35,36].

The deformation twins are also observed in the microstructure of fractured sample, as confirmed by the selected area electron diffraction (SAED) patterns (Figure 5f). It indicates that the accumulated dislocations inside grains are sufficient for the activation of deformation twins. The generation of deformation twinning normally leads to an increase in plasticity of steel during deformation, namely, so-called twinning induced plasticity (TWIP) effect. Similar to the effect of microbands, the austenite grains are divided into subregions by deformation twins, which significantly refine the grain size and promote the dynamic Hall-Petch effect. Furthermore, deformation twins can impede dislocation motion, playing a very important role in promoting the dislocation accumulation and causing a reduce in the dislocation mean free path, thereby contributing to stable the strain hardening [37].

For the studied steel deformed at the strain rate of 10^{-3} s^{-1}, dislocation slip is the major deformation mode at the early stages of straining, because the flow stress is below the critical stress for twinning. With increasing strain, the continuous strain hardening is a result of microband induced plasticity (MBIP) gradually changing to microband/TWIP. The synergy of microband and twinning enhance the strengthening ability of the present steel. In addition to the thermal softening mentioned in Section 3.2, it can be inferred that the adiabatic heating at high strain rate leads to the increase in SFE and inhibits the TWIP effect, which also induces the decreases in tensile strength and work hardening rate at the high strain rate of 10^{-1} s^{-1}.

4. Conclusions

In this work, the effect of strain rate on the mechanical properties of fully austenitic Fe-30Mn-8Al-1.0C (wt.%) steel was investigated under uniaxial tension, and corresponding microstructures were examined by SEM and TEM. The following conclusions can be drawn:

(1) With the increase in strain rate, the yield strength increases from 301 MPa to 381 MPa, and the ultimate tensile strength reaches the maximum of 801 MPa at 10^{-2} s^{-1}. The recovery of work hardening rate of studied steel is the most remarkable at the strain rate of 10^{-3} s^{-1}, and its elongation reaches as high as 72%. The deterioration of its mechanical properties at 10^{-1} s^{-1} might be related to thermal softening effect and the inhibition of TWIP effect.

(2) During the tensile deformation at the strain rate of 10^{-3} s^{-1}, dislocation arrays, Taylor lattices, microbands and deformation twins can be observed in sequence with increasing strain. This indicates that the continuous strain hardening results from both MBIP and TWIP, and a good combination of strength and ductility is thus achieved.

(3) Adiabatic heating that leads to the increase in SFE and inhibits the TWIP effect, as well as thermal softening occurring at the high strain rate of 10^{-1} s^{-1} jointly induces an anomalous decrease in tensile strength at such a high strain rate.

Author Contributions: Conceptualization, X.L. and P.C.; methodology, X.L. and P.C.; validation, J.D.; formal analysis, J.D.; investigation, J.D., X.G., J.C., C.L. and Q.P.; resources, X.L. and P.C.; writing—original draft preparation, J.D. and P.C.; writing—review and editing, X.L., P.C. and J.D.; visualization, X.L., P.C. and J.D.; supervision, X.L. and P.C.; project administration, J.D. and P.C.; funding acquisition, X.L. and P.C. All authors have read and agreed to the published version of the manuscript.

Funding: This research was funded by National Natural Science Foundation of China under grant numbers 52171108 and 51804072, and also by Fundamental Research Funds for the Central University under grant number N2202007.

Data Availability Statement: The raw/processed data required to reproduce these findings cannot be shared at this time as the data also forms part of an ongoing study.

Acknowledgments: Special thanks are also due to the instrumental or data analysis from Analytical and Testing Center, Northeastern University, China.

Conflicts of Interest: The authors declare no conflict of interest.

References

1. Park, K.T. Tensile deformation of low-density Fe–Mn–Al–C austenitic steels at ambient temperature. *Scr. Mater.* **2013**, *68*, 375–379. [CrossRef]
2. Sutou, Y.; Kamiya, N.; Umino, R.; Ohnuma, I.; Ishida, K. High-strength Fe–20Mn–Al–C-based Alloys with Low Density. *ISIJ Int.* **2010**, *50*, 893–899. [CrossRef]
3. Choo, W.K.; Kim, J.H.; Yoon, J.C. Microstructural change in austenitic Fe-30.0wt%Mn-7.8wt%Al-1.3wt%C initiated by spinodal decomposition and its influence on mechanical properties. *Acta Mater.* **1997**, *45*, 4877–4885. [CrossRef]
4. Frommeyer, G.; Brüx, U. Microstructures and Mechanical Properties of High-Strength Fe-Mn-Al-C Light-Weight TRIPLEX Steels. *Steel Res. Int.* **2006**, *77*, 627–633. [CrossRef]
5. Gutierrez-Urrutia, I.; Raabe, D. High strength and ductile low density austenitic FeMnAlC steels: Simplex and alloys strengthened by nanoscale ordered carbides. *Mater. Sci. Technol.* **2014**, *30*, 1099–1104. [CrossRef]
6. Chen, S.; Rana, R.; Haldar, A.; Ray, R.K. Current state of Fe-Mn-Al-C low density steels. *Prog. Mater. Sci.* **2017**, *89*, 345–391. [CrossRef]
7. Yao, M.J.; Dey, P.; Seol, J.-B.; Choi, P.; Herbig, M.; Marceau, R.K.W.; Hickel, T.; Neugebauer, J.; Raabe, D. Combined atom probe tomography and density functional theory investigation of the Al off-stoichiometry of κ-carbides in an austenitic Fe–Mn–Al–C low density steel. *Acta Mater.* **2016**, *106*, 229–238. [CrossRef]
8. Ren, P.; Chen, X.P.; Cao, Z.X.; Mei, L.; Li, W.J.; Cao, W.Q.; Liu, Q. Synergistic strengthening effect induced ultrahigh yield strength in lightweight Fe30Mn11Al-1.2C steel. *Mater. Sci. Eng. A* **2019**, *752*, 160–166. [CrossRef]
9. Moon, J.; Park, S.-J.; Jang, J.H.; Lee, T.-H.; Lee, C.-H.; Hong, H.-U.; Han, H.N.; Lee, J.; Lee, B.H.; Lee, C. Investigations of the microstructure evolution and tensile deformation behavior of austenitic Fe-Mn-Al-C lightweight steels and the effect of Mo addition. *Acta Mater.* **2018**, *147*, 226–235. [CrossRef]
10. Yoo, J.D.; Hwang, S.W.; Park, K.-T. Factors influencing the tensile behavior of a Fe–28Mn–9Al–0.8C steel. *Mater. Sci. Eng. A* **2009**, *508*, 234–240. [CrossRef]
11. Yang, H.K.; Zhang, Z.J.; Dong, F.Y.; Duan, Q.Q.; Zhang, Z.F. Strain rate effects on tensile deformation behaviors for Fe–22Mn–0.6C–(1.5Al) twinning-induced plasticity steel. *Mater. Sci. Eng. A* **2014**, *607*, 551–558. [CrossRef]
12. Leslie, W.C.; Rauch, G.C. Precipitation of carbides in low-carbon Fe-Al-C alloys. *Metall. Mater. Trans. A* **1987**, *9*, 343–349. [CrossRef]
13. Talonen, J.; Hänninen, H.; Nenonen, P.; Pape, G. Effect of strain rate on the strain-induced γ → α'-martensite transformation and mechanical properties of austenitic stainless steels. *Metall. Mater. Trans. A* **2005**, *36*, 421–432. [CrossRef]
14. Kundu, A.; Chakraborti, P.C. Effect of strain rate on quasistatic tensile flow behaviour of solution annealed 304 austenitic stainless steel at room temperature. *J. Mater. Sci.* **2010**, *45*, 5482–5489. [CrossRef]
15. Curtze, S.; Kuokkala, V.-T. Dependence of tensile deformation behavior of TWIP steels on stacking fault energy, temperature and strain rate. *Acta Mater.* **2010**, *58*, 5129–5141. [CrossRef]

16. Gerold, V.; Karnthaler, H.P. On the origin of planar slip in f.c.c. alloys. *Acta Metall.* **1989**, *37*, 2177–2183. [CrossRef]
17. Han, D.; Zhang, Y.J.; Li, X.W. A crucial impact of short-range ordering on the cyclic deformation and damage behavior of face-centered cubic alloys: A case study on Cu-Mn alloys. *Acta Mater.* **2021**, *205*, 116559. [CrossRef]
18. Liu, R.T. *Failure Analysis of Mechanical Parts*, 1st ed.; Harbin Institute of Technology Press: Harbin, China, 2003; pp. 103–104.
19. Wang, R.; Lu, C.; Tieu, K.A.; Gazder, A.A. Slip system activity and lattice rotation in polycrystalline copper during uniaxial tension. *J. Mater. Res. Technol.* **2022**, *18*, 508–519. [CrossRef]
20. Kim, C.; Terner, M.; Hong, H.-U.; Lee, C.-H.; Park, S.-J.; Moon, J. Influence of inter/intra-granular κ-carbides on the deformation mechanism in lightweight Fe-20Mn-11.5Al-1.2C steel. *Mater. Charact.* **2020**, *161*, 110142. [CrossRef]
21. Liu, J.; Wu, H.; He, J.; Yang, S.; Ding, C. Effect of κ-carbides on the mechanical properties and superparamagnetism of Fe–28Mn–11Al-1.5/1.7C–5Cr lightweight steels. *Mater. Sci. Eng. A* **2022**, *849*, 143462. [CrossRef]
22. Cheng, W.-C.; Cheng, C.-Y.; Hsu, C.-W.; Laughlin, D.E. Phase transformation of the L1$_2$ phase to kappa-carbide after spinodal decomposition and ordering in an Fe–C–Mn–Al austenitic steel. *Mater. Sci. Eng. A* **2015**, *642*, 128–135. [CrossRef]
23. Chen, P.; Fu, J.; Xu, X.; Lin, C.; Pang, J.C.; Li, X.W.; Misra, R.D.K.; Wang, G.D.; Yi, H.L. A high specific Young's modulus steel reinforced by spheroidal kappa-carbide. *Mater. Sci. Technol.* **2021**, *87*, 54–59. [CrossRef]
24. Chen, P.; Xiong, X.C.; Wang, G.D.; Yi, H.L. The origin of the brittleness of high aluminum pearlite and the method for improving ductility. *Scr. Mater.* **2016**, *124*, 42–46. [CrossRef]
25. Chen, P.; Li, X.; Yi, H. The κ-Carbides in Low-Density Fe-Mn-Al-C Steels: A Review on Their Structure, Precipitation and Deformation Mechanism. *Metals* **2020**, *10*, 1021. [CrossRef]
26. Yao, M.J.; Welsch, E.; Ponge, D.; Haghighat, S.M.H.; Sandlöbes, S.; Choi, P.; Herbig, M.; Bleskov, I.; Hickel, T.; Lipinska-Chwalek, M.; et al. Strengthening and strain hardening mechanisms in a precipitation-hardened high-Mn lightweight steel. *Acta Mater.* **2017**, *140*, 258–273. [CrossRef]
27. Kimura, Y.; Handa, K.; Hayashi, K.; Mishima, Y. Microstructure control and ductility improvement of the two-phase γ-Fe/κ-(Fe,Mn)$_3$AlC alloys in the Fe–Mn–Al–C quaternary system. *Intermetallics* **2004**, *12*, 607–617. [CrossRef]
28. Zhang, Y.J.; Han, D.; Li, X.W. A unique two-stage strength-ductility match in low solid-solution hardening Ni-Cr alloys: Decisive role of short range ordering. *Scr. Mater.* **2020**, *178*, 269–273. [CrossRef]
29. Han, D.; Wang, Z.Y.; Yan, Y.; Shi, F.; Li, X.W. A good strength-ductility match in Cu-Mn alloys with high stacking fault energies: Determinant effect of short range ordering. *Scr. Mater.* **2017**, *133*, 59–64. [CrossRef]
30. Gutierrez-Urrutia, I.; Raabe, D. Multistage strain hardening through dislocation substructure and twinning in a high strength and ductile weight-reduced Fe–Mn–Al–C steel. *Acta Mater.* **2012**, *60*, 5791–5802. [CrossRef]
31. Kuhlmann-Wilsdorf, D. Theory of plastic deformation:—properties of low energy dislocation structures. *Mater. Sci. Eng. A* **1989**, *113*, 1–41. [CrossRef]
32. Canadinc, D.; Sehitoglu, H.; Maier, H.J.; Niklasch, D.; Chumlyakov, Y.I. Orientation evolution in Hadfield steel single crystals under combined slip and twinning. *Int. J. Sol. Struct.* **2007**, *44*, 34–50. [CrossRef]
33. Ma, B.; Li, C.; Zheng, J.; Song, Y.; Han, Y. Strain hardening behavior and deformation substructure of Fe–20/27Mn–4Al–0.3C non-magnetic steels. *Mater. Des.* **2016**, *92*, 313–321. [CrossRef]
34. Yoo, J.D.; Hwang, S.W.; Park, K.-T. Origin of Extended Tensile Ductility of a Fe-28Mn-10Al-1C Steel. *Metall. Mater. Trans. A* **2009**, *40*, 1520–1523. [CrossRef]
35. Yoo, J.D.; Park, K.-T. Microband-induced plasticity in a high Mn–Al–C light steel. *Mater. Sci. Eng. A* **2008**, *496*, 417–424. [CrossRef]
36. Liu, X.; Wu, Y.; Wang, Y.; Chen, J.; Bai, R.; Gao, L.; Xu, Z.; Wang, W.Y.; Tan, C.; Hui, X. Enhanced dynamic deformability and strengthening effect via twinning and microbanding in high density NiCoFeCrMoW high-entropy alloys. *J. Mater. Sci. Technol.* **2022**, *127*, 164–176. [CrossRef]
37. Yao, K.; Min, X. Static and dynamic Hall–Petch relations in {332}<113> TWIP Ti–15Mo alloy. *Mater. Sci. Eng. A* **2021**, *827*, 142044. [CrossRef]

Article

Quenching Stress of Hot-Rolled Seamless Steel Tubes under Different Cooling Intensities Based on Simulation

Zhenlei Li [1], Rui Zhang [1], Dong Chen [1], Qian Xie [2,*], Jian Kang [1], Guo Yuan [1,*] and Guodong Wang [1]

1 State Key Laboratory of Rolling and Automation, Northeastern University, Shenyang 110819, China
2 School of Metallurgic Engineering, Anhui University of Technology, Ma'anshan 243002, China
* Correspondence: xieqian@ahut.edu.cn (Q.X.); yuanguoneu@163.com (G.Y.)

Abstract: Large residual stress occurs during the quenching process of hot-rolled seamless steel tubes, which results in bending, cracking, and ellipticity exceeding standards and seriously affects the quality of hot-rolled seamless steel tubes. In addition, the stress generation mechanism of hot-rolled seamless steel tubes is different from that of steel plates due to the characteristics of annular section. In this research, the finite element simulation method was used to study the quenching residual stress of seamless steel tubes with different cooling intensities. The variation law of temperature and stress on the steel tube under different cooling intensities were analyzed. The results show that the radial stress was close to 0, and the circumferential and axial stresses were the main factors affecting the quality of the steel tube. With the increase in the cooling time, the magnitude and direction of each stress component of the steel tube changed simultaneously. Finally, a typical stress distribution state of "external compressive stress, internal tensile stress" was presented in the thickness direction of the steel tube. Furthermore, with the increase in the cooling intensity, the residual stress of the steel tube gradually increased and was mainly concentrated on the near wall of the steel tube.

Keywords: seamless steel tubes; residual stress; finite element simulation; heat transfer coefficient; cooling intensity

1. Introduction

As an important steel variety, hot-rolled seamless steel tubes are widely used in energy extraction, petrochemical and machinery manufacturing, and other fields [1–3]. With the acceleration of economic construction and the increasingly complex environment for energy extraction, such as oil and natural gas, higher performance requirements must be met by hot-rolled seamless steel tubes. As one of the important links in the production process of metal materials, the heat treatment process can fully tap the potential of the material, increase intensity, and improve the plasticity and welding performance [4]. However, during the quenching process of hot-rolled seamless steel tubes, due to the cooling difference between the inner and outer walls, there is a large temperature gradient in thickness, which leads to internal thermal stress. When the stress exceeds the yield intensity of the steel tube, problems occur, such as bending, cracking, or ellipticity changes, which seriously affect product quality [5,6]. In addition, the distribution of residual stress is greatly related to the shape. Compared with the plates, the characteristics of the annular hollow section of hot-rolled seamless steel tubes make the stress generation mechanism more complicated, which greatly increases the difficulty of residual stress analysis. Thus, the research progress in quenching stress in the field of steel tubes is far behind that of steel plates. It is important to study the generation mechanism and evolution law of stress in the quenching process of hot-rolled seamless steel tubes, which greatly contribute to the improvement in the quality of hot-rolled seamless steel tubes.

Previous studies mainly focused on the heat transfer of steel tubes or cylindrical surfaces [7–11], but stress changes during cooling were rarely reported. Ali et al. [12] studied

Citation: Li, Z.; Zhang, R.; Chen, D.; Xie, Q.; Kang, J.; Yuan, G.; Wang, G. Quenching Stress of Hot-Rolled Seamless Steel Tubes under Different Cooling Intensities Based on Simulation. *Metals* **2022**, *12*, 1363. https://doi.org/10.3390/met12081363

Academic Editors: Giovanni Meneghetti and Denis Benasciutti

Received: 5 July 2022
Accepted: 13 August 2022
Published: 16 August 2022

Publisher's Note: MDPI stays neutral with regard to jurisdictional claims in published maps and institutional affiliations.

Copyright: © 2022 by the authors. Licensee MDPI, Basel, Switzerland. This article is an open access article distributed under the terms and conditions of the Creative Commons Attribution (CC BY) license (https://creativecommons.org/licenses/by/4.0/).

the comprehensive effects of time, material properties, and the ratio of the inner and outer radius on the transient temperature gradient and provided an empirical solution for the thermal stress of the cylinder under the conditions of heating, secondary heating, and forced air impact cooling. Schemmel et al. [13] investigated the formation of residual stress and the evolution of phase components during the quenching process of cylindrical specimens with different sizes. They found that the surface residual stress changed from tensile stress to compressive stress when the size of the steel tube was increased. Hata et al. [14] theoretically analyzed the thermal stress induced by thermal shock and the stress-focusing effect induced by phase transformation stress. They pointed out that, in the quenched state, cracks may occur in the center of the cylindrical rod due to the interaction of the thermal phase transition stress-focusing effect and the phase transition stress-focusing effect. Oliveira et al. [15] showed the thermodynamic behavior of a cylinder and established a non-isothermal multiphase constitutive equation. Chen et al. [16] carried out related research on the problem of axial cracks caused by thermal shock in coated hollow cylinders; they used the finite element method to calculate the transient temperature and induced the thermal stress and crack tip stress intensity factors of the inner surface of the cylinder after convective cooling. Devynck et al. [17] studied the effect of boiling heat transfer and phase transformation on the quenching deformation of steel tubes by simulation and experimentation. Their calculated results regarding the bending of the steel tube corresponded to the measured values. Leitner et al. [18] used finite element simulation and experimentation to study how to use a controlled cooling strategy to control residual stress and phase structure in multiphase steel tubes, and pointed out that using a low cooling rate would result in a lower plasticity and residual stress. Yang et al. [19] also used the finite element method to study the effects of quenching water temperature, the rod length-to-diameter ratio, pre-stretching ratio, and stretching rate on the residual stress of quenched rods after cold stretching. They showed that the maximum residual tensile and compressive stress of the quenched rod decreases with the increase in the water temperature.

The magnitude of residual stress of hot-rolled seamless steel tubes depends on the cooling intensity during the quenching process. At present, the cooling intensity control of steel tubes can be realized by changing the water flow and temperature of the cooling medium [20]. In this paper, relevant research was carried out on the variation law of temperature and stress during the quenching process of hot-rolled seamless steel tubes under different cooling intensities. The purpose was to explore the inherent laws of cooling intensity, temperature field, and stress field and to provide data support and theoretical reference for the design of a heat treatment process for eliminating the residual stress of seamless steel tubes.

2. FEM Model of Quenching Process

2.1. Materials and Methods

As shown in Figure 1, the heat treatment process of materials is a complex process of chemical composition, temperature field, metallurgy field, and stress field coupling [21–24]. The metal workpiece after heat treatment can undergo a solid-phase transformation when the temperature changes in the solid-state range. Different phases can be obtained by controlling the progress of the solid-state-phase transition. There are differences in the stacking ratio and specific heat capacity of different phases. During the cooling process of the workpiece, due to the existence of the temperature gradient in the workpiece cross-section, the phase transformation of the workpiece cannot be carried out at the same time. This internal stress due to phase-transition asynchrony is called microstructure stress, and the resulting strain is called microstructure strain. It is difficult to obtain the real-time changing state of stress and predict the residual stress field. Thus, it is difficult to carry out targeted prevention and elimination measures. As a finite element simulation software, ANSYS can handle any material, any complex shape, any boundary, and any time or heat treatment process, which can shorten the research time and lower the research and development costs. Therefore, this paper used ANSYS software to simulate the temperature field and stress field of hot-rolled seamless steel tubes under different cooling intensities.

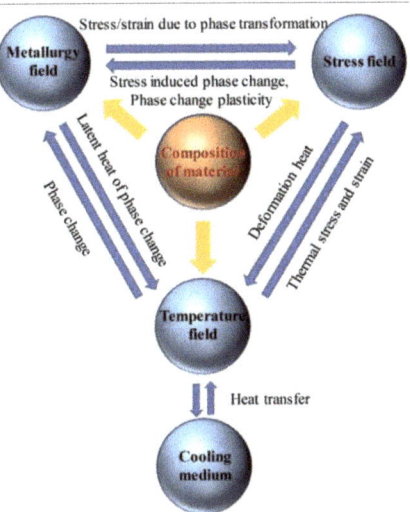

Figure 1. Multiphysics and coupling interactions involved in heat treatment engineering.

The size of the steel tube was 140 × 20 × 400 mm. Because the steel tube was an axisymmetric model, the 1/8 model was selected for simulation in order to reduce the simulation time, as shown in Figure 2. The material of the steel tube that was to be heat-treated was 310S stainless steel, which experiences no phase transformation process during heat treatment. Thus, the research problem was simplified to the interaction between the temperature field and the stress field. The thermophysical parameters of 310S stainless steel were determined through linear interpolation, as shown in Figure 3.

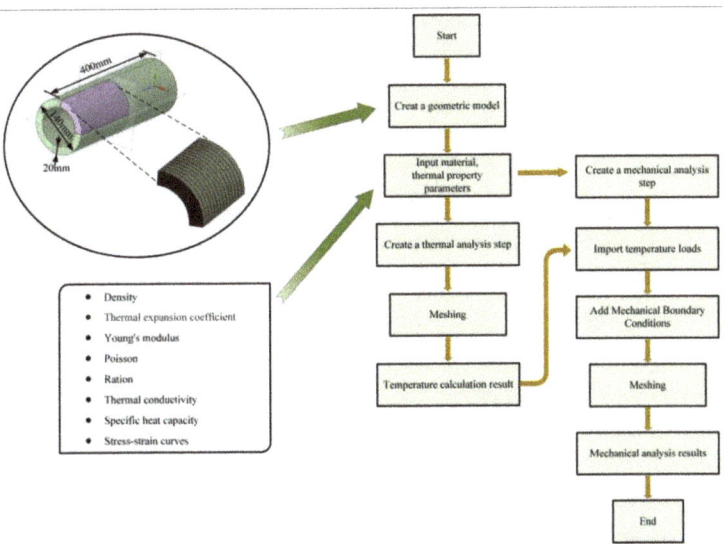

Figure 2. Finite element model and calculation process.

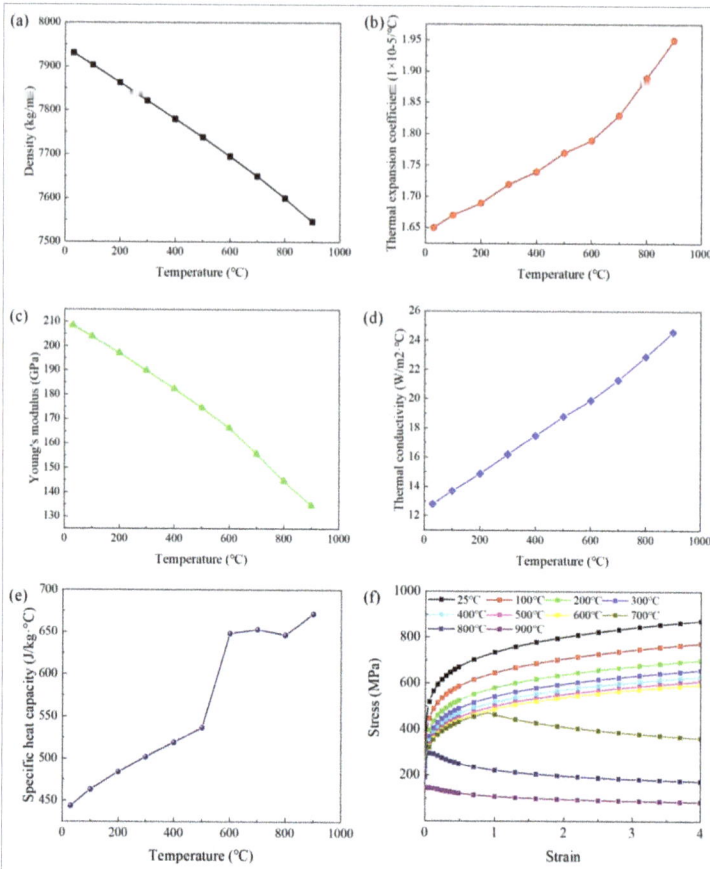

Figure 3. The thermophysical parameters of 310S stainless steel. (**a**) desnity; (**b**) thermal expansion cofficient; (**c**) Young's modulus; (**d**) thermal conducivity; (**e**) specific heat capacity; (**f**) stress−strain curves.

2.2. Mathematical Model

The calculation of the temperature field is achieved using the three-dimensional thermal conductivity differential equation in the cylindrical coordinate system, and its expression is shown as follows [7,10,25]:

$$\frac{1}{r}\frac{\partial}{\partial r}\left(\lambda r\frac{\partial T}{\partial r}\right) + \frac{1}{r^2}\frac{\partial}{\partial \varphi}\left(\lambda\frac{\partial T}{\partial \varphi}\right) + \frac{\partial}{\partial z}\left(\lambda\frac{\partial T}{\partial z}\right) + \dot{\Phi} = \rho c \frac{\partial T}{\partial t} \quad (1)$$

where λ is the thermal conductivity; ρ is the material density; c is the material-specific heat capacity; T is the temperature of the steel tube; t is the time; r is the radial distance of the steel tube; φ is the azimuth angle; z is the height; and $\dot{\Phi}$ is the internal heat source, which comes from the heat released by the phase transition during the quenching process, and because there is no phase transition in this simulation, the value is 0.

In order to determine the unique thermal conductivity differential equation, the initial and boundary conditions need to be given. The initial condition refers to the temperature distribution (T_0) of the object area at the initial time, which is expressed by Equation (2):

$$T|_{t=0} = T_0 \quad (2)$$

The third type of boundary condition is used to set the temperature of the cooling medium and the surface heat transfer coefficient during the quenching process of the steel tube, and its expression is shown as follows:

$$-\lambda \left(\frac{\partial T}{\partial n}\right) = h\left(T_w - T_f\right) \qquad (3)$$

In this finite element simulation, the initial temperature of the steel tube (T_w) is 900 °C, the cooling water temperature (T_f) is 30 °C. Using the single-variable method, only the cooling intensity of the single wall is changed each time. When the heat transfer coefficient of the inner wall (h_{inner}) is 600 W/(m²·°C), the heat transfer coefficient of the outer wall (h_{outer}) is set to four control groups: 1000, 2000, 3000, and 4000 W/(m²·°C). Similarly, when the heat transfer coefficient of the outer wall (h_{outer}) is 1000 W/(m²·°C), the heat transfer coefficient of the outer wall (h_{inner}) is set to four control groups: 200, 600, 1000, and 1400 W/(m²·°C). Since the cooling intensity of inner wall is lower than that of outer wall in actual production, the heat transfer coefficient of the inner wall is smaller than that of the outer wall.

For the linear elastic model, the incremental relationship between stress and strain is expressed in Equation (4).

$$d\sigma = D_e d\varepsilon \qquad (4)$$

$$D_e = \frac{E}{(1+v)(1-2v)} \begin{bmatrix} 1-v & v & v & 0 & 0 & 0 \\ v & 1-v & v & 0 & 0 & 0 \\ v & v & 1-v & 0 & 0 & 0 \\ 0 & 0 & 0 & \frac{1-2v}{2} & 0 & 0 \\ 0 & 0 & 0 & 0 & \frac{1-2v}{2} & 0 \\ 0 & 0 & 0 & 0 & 0 & \frac{1-2v}{2} \end{bmatrix} \qquad (5)$$

where $d\sigma$ is the stress increment, $d\varepsilon$ is the strain increment, D_e is the elastic modulus matrix, E is the elastic modulus, and v is the Poisson ratio.

Because stainless steel is used in this simulation, there is no phase transition, and the phase transition strain increment is 0. For the thermo-elastoplastic model for calculating thermal stress, the strain increment is expressed by Equation (6).

$$d\varepsilon = d\varepsilon_e + d\varepsilon_p + d\varepsilon_T \qquad (6)$$

where $d\varepsilon_e$, $d\varepsilon_p$, and $d\varepsilon_T$ are the elastic strain increment, the plastic strain increment, and the thermal strain increment.

$$d\varepsilon_e = \frac{1}{2G}d\sigma' + \frac{1-2v}{E}d\sigma_m \delta_{ij} \qquad (7)$$

where $d\sigma'$ and $d\sigma_m$ are the stress deviator increment and mean stress increment.

The shear modulus G is shown in the following equation:

$$G = \frac{E}{2(1+v)} \qquad (8)$$

The Kronecker symbol is defined as follows:

$$\delta_{ij} = \begin{cases} 1, & i = j \\ 0, & i \neq j \end{cases} \qquad (9)$$

The flow criterion is an assumption that indicates the direction of the plastic deformation increment after the material reaches yield, that is, a proportional relationship between the components of the plastic deformation increment. The plastic strain increment is shown in the following equation:

$$d\varepsilon_p = d\lambda \sigma' = \frac{3}{2}\frac{d\overline{\varepsilon_p}}{\sigma}\sigma' \tag{10}$$

where $d\lambda$, σ', $d\overline{\varepsilon_p}$, and σ are the instantaneous scaling factor, the stress deviator, the equivalent plastic strain increment, and the equivalent stress.

The thermal strain increment can be expressed as:

$$d\varepsilon_T = \sum_{k=1}^{n} y_k \alpha_k dT \tag{11}$$

In the formula, y_k refers to the volume fraction of the phase k in the material; n refers to number of phases in the material; and α_k refers to the thermal expansion coefficient of the phase k related to the temperature. Since 310S stainless steel is used in this paper, only the austenite phase exists, so the thermal strain increment formula can be simplified as:

$$d\varepsilon_T = \alpha dT \tag{12}$$

2.3. Verification of Other Finite Element Simulation Cases

In order to verify the accuracy of the residual stress simulation, the finite element simulation results are compared with the experimental data in reference [26]. Reference [26] studied the quenching residual stress distribution of 7050 Al ingot by means of experiments and finite element simulation. Because the ingot is a symmetrical model, in order to reduce the calculation time, only a quarter of the model is used in this simulation. As shown in Figure 4, the simulation data gathered in this paper are compared with the residual stress in reference [26]. It can be seen from the figure that although a certain error occur in the simulated data compared to the experimental data, the same trend is maintained.

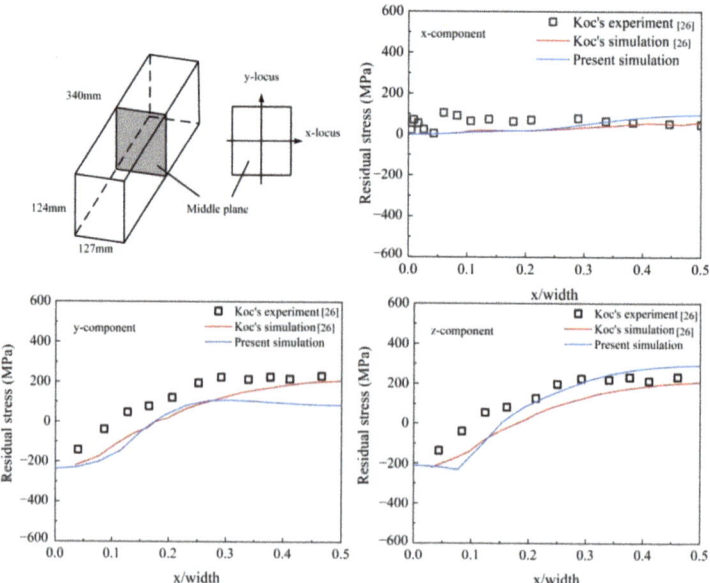

Figure 4. Comparison of the simulated data and experimental data [26].

3. Results and Discussion

3.1. Different Cooling Intensities of the Outer Wall

Figure 5 shows the temperature curves of the steel tube under different cooling intensities of the outer wall. It can be seen from Equation (3) that the heat transfer was related to the temperature gradient. At the initial moment, the temperature of the steel tube was high, there was a large temperature gradient within the surrounding environment, the heat transfer rate was fast, and the temperature dropped rapidly. Then, as the cooling time increased, the temperature of the steel tube gradually decreased. The gradient, with respect to the ambient temperature, gradually decreased, and the decreasing trend of the temperature gradually slowed down. Eventually, the steel tube cooled to ambient temperature with zero heat transfer. By comparison, it was found that with the increase in the convection heat transfer coefficient of the outer wall, the temperature of the outer wall decreased faster. Figure 6 shows the temperature distribution cloud map of the steel tube.

Figure 7 shows the equivalent stress under different cooling intensities of the outer wall. The von Mises criterion was adopted as the effective stress. It was calculated by Equation (13). There was no sign of the calculation result, so the state of stress could not be judged.

$$(\sigma_r - \sigma_\theta)^2 + (\sigma_\theta - \sigma_z)^2 + (\sigma_z - \sigma_r)^2 = 2\sigma_s^2 \qquad (13)$$

The maximum equivalent stress was 341.99 MPa when h_{outer} = 1000 W/(m²·°C), as shown in Figure 7a. When h_{outer} = 2000 W/(m²·°C), the maximum equivalent stress was 397.42 MPa and increased by 55.43 MPa. When h_{outer} = 3000 W/(m²·°C), the maximum equivalent stress was 428.01 MPa and increased by 30.59 MPa. When h_{outer} = 4000 W/(m²·°C), the maximum equivalent stress was 447.69 MP and increased by 19.68 MPa. The maximum equivalent stress was mainly concentrated on the outer wall. With the increase in the cooling intensity, the maximum equivalent stress of the steel tube gradually increased, but the increase trend gradually slowed down.

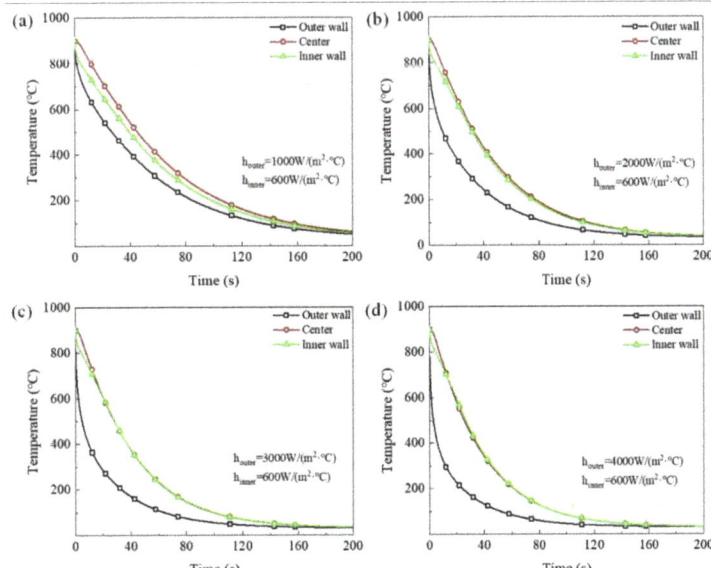

Figure 5. Temperature curves of the steel tube at different cooling intensities of the outer wall. (**a**) h_{outer} = 1000 W/(m²·°C); (**b**) h_{outer} = 2000 W/(m²·°C); (**c**) h_{outer} = 3000 W/(m²·°C); (**d**) h_{outer} = 4000 W/(m²·°C).

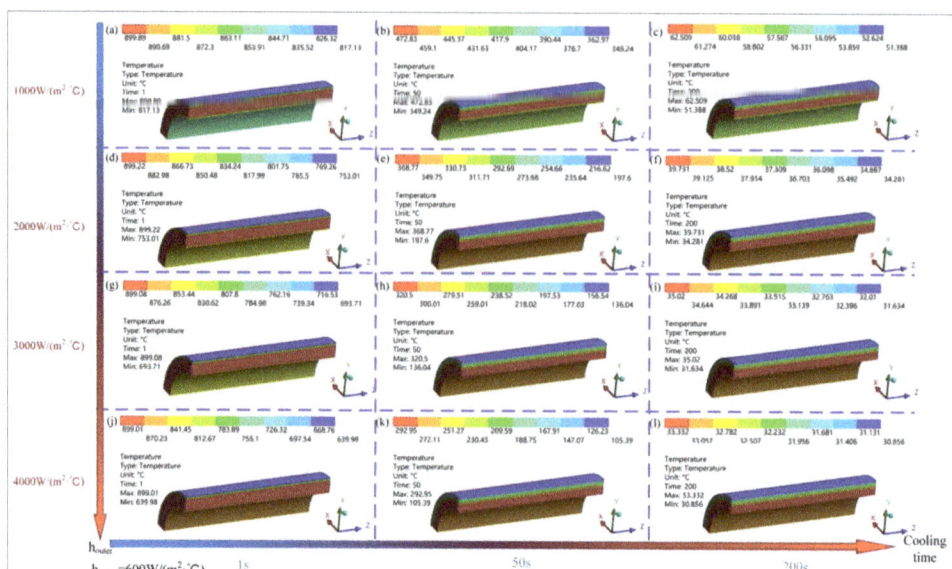

Figure 6. The temperature fields' evolution in the quenching process of the steel tube under different cooling intensities of the outer wall. (**a**) h_{outer} = 1000 W/(m²·°C), t = 1 s; (**b**) h_{outer} = 1000 W/(m²·°C), t = 50 s; (**c**) h_{outer} = 1000 W/(m²·°C), t = 200 s; (**d**) h_{outer} = 2000 W/(m²·°C), t = 1 s; (**e**) h_{outer} = 2000 W/(m²·°C), t = 50 s; (**f**) h_{outer} = 2000 W/(m²·°C), t = 200 s; (**g**) h_{outer} = 3000 W/(m²·°C), t = 1 s; (**h**) h_{outer} = 3000 W/(m²·°C), t = 50 s; (**i**) h_{outer} = 3000 W/(m²·°C), t = 200 s; (**j**) h_{outer} = 4000 W/(m²·°C), t = 1 s; (**k**) h_{outer} = 4000 W/(m²·°C), t = 50 s; (**l**) h_{outer} = 4000 W/(m²·°C), t = 200 s.

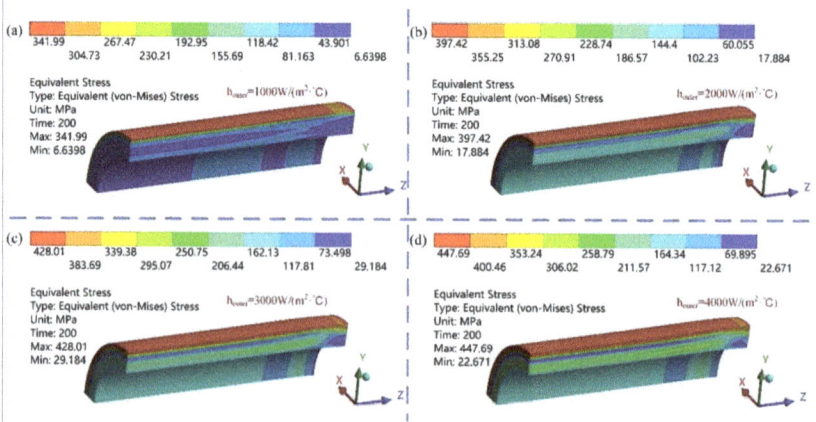

Figure 7. Distribution of the equivalent stress filed at different cooling intensities of the outer wall. (**a**) h_{outer} = 1000 W/(m²·°C); (**b**) h_{outer} = 2000 W/(m²·°C); (**c**) h_{outer} = 3000 W/(m²·°C); (**d**) h_{outer} = 4000 W/(m²·°C).

Figure 8 shows the variation curves of the radial stress, circumferential stress, and axial stress with the time at different positions. As can be seen from Figure 8, the radial stress (σ_r) was close to 0, which was much smaller than the circumferential (σ_θ) and axial stress (σ_z). The circumferential and axial stresses were the main factors affecting the quality of the steel tube during the quenching process. With the increase in the cooling intensity of the outer wall, the stress components of the steel tube increased. In the early stage of cooling of the steel tube, the outer wall rapidly cooled, forming a large temperature gradient with the inside and shrinking to the inside. Each stress component reached its peak in a short time. With the increase in the cooling time, each stress component gradually decreased and the direction changed.

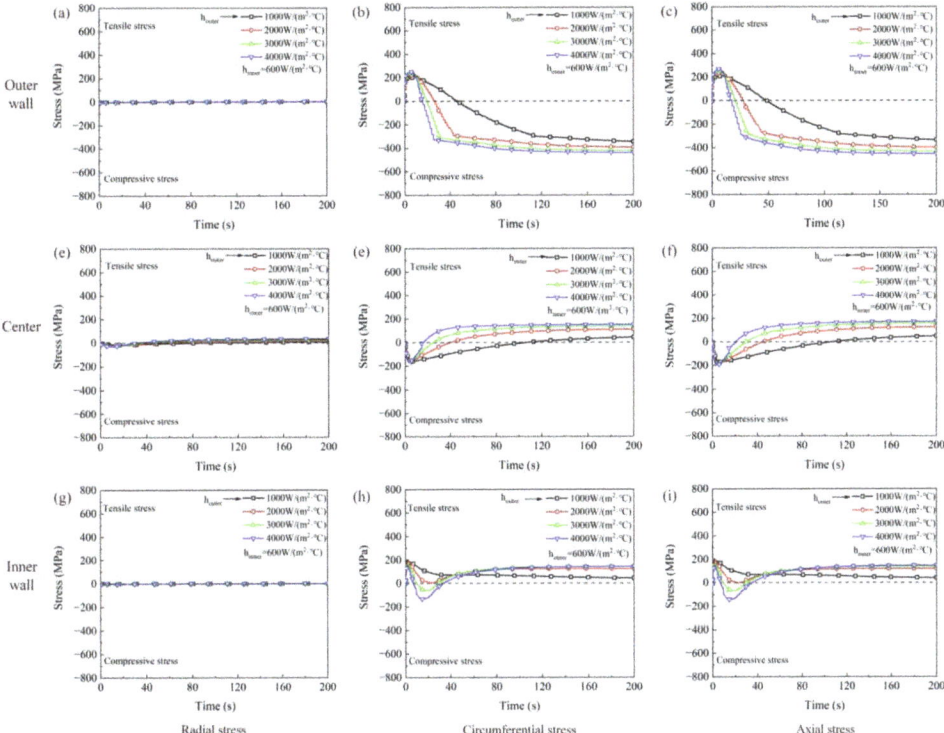

Figure 8. Stress-time curves of different positions under different cooling intensities of the outer wall. (**a**) radial stress of the outer wall; (**b**) circumferential stress of the outer wall; (**c**) axial stress of the outer wall; (**d**) radial stress of the center; (**e**) circumferential stress of the center; (**f**) axial stress of the center; (**g**) radial stress of the inner wall; (**h**) circumferential stress of the inner wall; (**i**) axial stress of the inner wall.

As shown in Figure 8b,c, the outer wall of the steel tube was restricted by the internal material, which hindered the further development of its shrinkage. The outer wall was subjected to tensile stress in the circumferential and axial directions. As the temperature of the outer wall further decreased, the cooling rate of the outer wall was lower than the cooling rate of the center, so that the shrinkage of the center was greater than that of the outer wall, and the stress on the outer wall gradually decreased. When the steel tube was cooled to a certain temperature, the direction of the stress changed. The outer wall was subjected to compressive stress in the circumferential and axial directions. On the contrary, as shown in Figure 8e,f, the center was subjected to pressure from the outer wall at the initial stage; as the cooling rate of the outer wall gradually decreased, the compressive stress on the center gradually decreased, and the direction of the stress finally changed and became tensile stress.

It can be seen from Figure 8h,i that the direction of the stress on the inner wall changed twice. This was due to a shift in the magnitude relationship between the cooling rates of the outer wall, the center, and the inner wall. As shown in Figure 8, in the initial cooling stage, the cooling rate of the inner wall of the steel tube was greater than the heat conduction of the wall, and the inner wall surface shrank, so the inner wall was subjected to the tensile stress exerted by the center in the circumferential and axial directions. While the cooling time increased, the cooling rate of the center was greater than that of the inner wall. The tensile stress on the inner wall gradually decreased, and the direction of the stress changed. When the cooling rate of both the inner wall and the center was greater than that of the outer wall surface, the direction of stress changed again and the inner wall was subjected to the tensile stress exerted by the outer wall.

In the radial direction, there was tensile stress at the center, and the maximum stress appeared near the outer wall, while the stress near the inner and outer walls was close to 0, as seen in Figure 9. In the axial direction, there was compressive stress on the surface and tensile stress at the center, and the maximum stress appeared on the outer wall of the steel tube. The distribution of the circumferential stress was similar to that of the axial stress. The magnitude of the residual stress increased with the increase in the cooling intensity, and the position where the stress direction changed gradually moved to the center, with an increase in the cooling intensity on the outer wall.

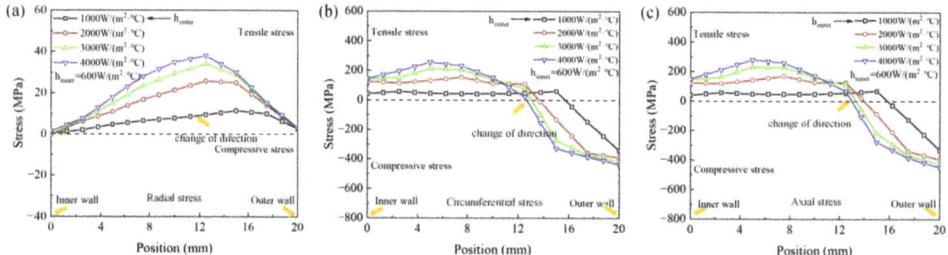

Figure 9. Stress distribution curve under different cooling intensities of the outer wall. (**a**) radial stress; (**b**) circumferential stress; (**c**) axial stress.

3.2. Different Cooling Intensities of the Inner Wall

With the increase in the cooling intensity of the inner wall, the cooling rate gradually increased, and the temperature curves of the inner and outer walls gradually tended to be consistent (as is shown in Figure 10). When the cooling intensity of the inner and outer walls were the same, the cooling rate of the inner wall was smaller than the cooling rate of the outer wall. This is because the heat transfer area gradually increased from the inside to the outside during the cooling process of the steel tube. On the contrary, the heat transfer area gradually decreased from the outside to the inside. Figure 11 shows the cloud map of the temperature distribution under different cooling intensities of the inner wall. As the cooling intensity of the inner wall increased, the position of the highest temperature gradually moved to the center of the steel tube.

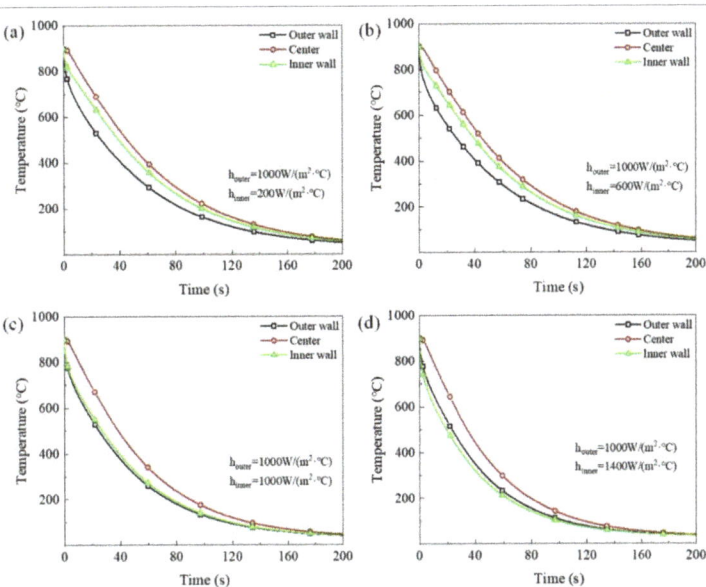

Figure 10. Temperature curves of the steel tube under different cooling intensities of the inner wall. (a) h_{inner} = 200 W/(m²·°C); (b) h_{inner} = 600 W/(m²·°C); (c) h_{inner} = 1000 W/(m²·°C); (d) h_{inner} =1400 W/(m²·°C).

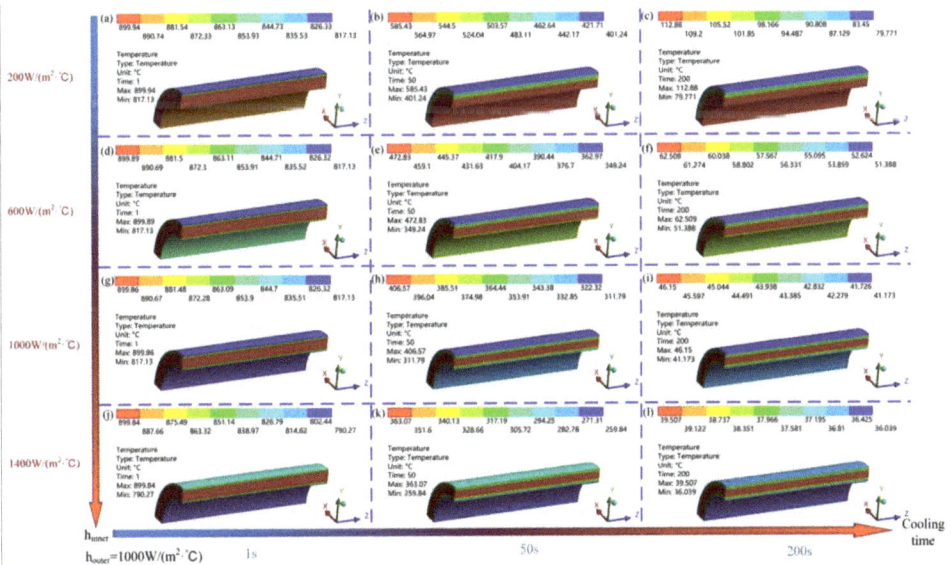

Figure 11. The temperature fields' evolution in the quenching process of the steel tube under different cooling intensities of the inner wall. (a) h_{inner} = 200 W/(m²·°C), t = 1 s; (b) h_{inner} = 200 W/(m²·°C), t = 50 s; (c) h_{inner} = 200 W/(m²·°C), t = 200 s; (d) h_{inner} = 600 W/(m²·°C), t = 1 s; (e) h_{inner} = 600 W/(m²·°C), t = 50 s; (f) h_{inner} = 600 W/(m²·°C), t = 200 s; (g) h_{inner} = 1000 W/(m²·°C), t = 1 s; (h) h_{inner} = 1000 W/(m²·°C), t = 50 s; (i) h_{inner} = 1000 W/(m²·°C), t = 200 s; (j) h_{inner} = 1400 W/(m²·°C), t = 1 s; (k) h_{inner} = 1400 W/(m²·°C), t = 50 s; (l) h_{inner} = 1400 W/(m²·°C), t = 200 s.

Figure 12 shows the cloud diagram of the equivalent stress at different cooling intensities of the inner wall. When the cooling intensity of the inner wall was small, the maximum equivalent stress was mainly concentrated on the outer wall. However, when the cooling intensity of the inner wall was greater than the cooling intensity of the outer wall, the maximum equivalent stress was located on the inner wall. With the increase in the cooling intensity of the inner wall, the maximum equivalent stress first decreased and then increased.

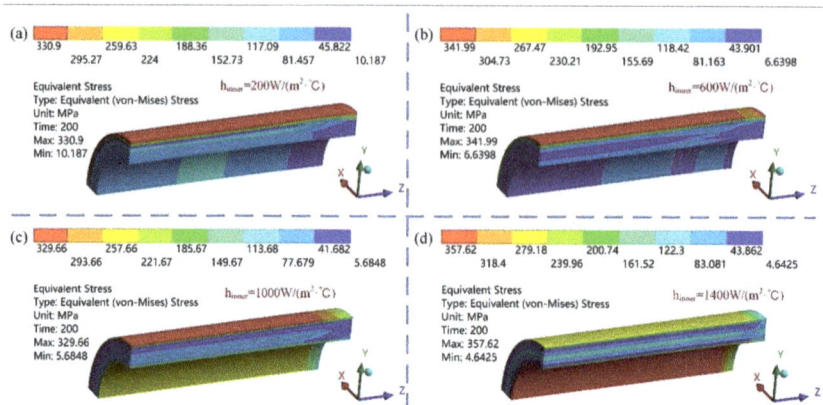

Figure 12. Distribution of the equivalent stress field at different cooling intensities of the inner wall. (**a**) h_{inner} = 200 W/(m²·°C); (**b**) h_{inner} = 600 W/(m²·°C); (**c**) h_{inner} = 1000 W/(m²·°C); (**d**) h_{inner} = 1400 W/(m²·°C).

The initial moment of circumferential and axial stresses on the outer wall was tensile stress, as shown in Figure 13. With the increase in the cooling time, the stress state changed to compressive stress. The stress at the center changed from compressive stress to tensile stress. This was similar to the change law of stress when changing the cooling intensity of the outer wall. It is worth noting that when the cooling intensity of the inner wall was 200 W/(m²·°C), although the initial stress in the circumferential and axial directions was tensile stress, it transformed into compressive stress in a short period of time, and transformed into tensile stress with the increasing cooling time. As the cooling intensity of the inner wall increased, the circumferential and axial stresses of the inner wall only changed the direction of the stress once, from tensile stress to compressive stress.

Figure 14 shows the distribution of residual stress in the direction of the steel tube at different inner wall cooling intensities. When the cooling intensity of the inner wall was less than that of the outer wall, all the radial stresses were tensile stress. When the cooling intensity of the inner wall was greater than that of the outer wall, the stress of the near outer wall was tensile stress and the stress of the near inner wall was compressive stress. With the increase in the cooling intensity, the circumferential and axial stress changed from tensile stress to compressive stress, and the position of transformation moved gradually closer to the center. The position where the tensile stress changed from compressive stress was close to the outer wall, which was contrary to the changing trend of the cooling intensity of the outer wall, as seen in Figure 9.

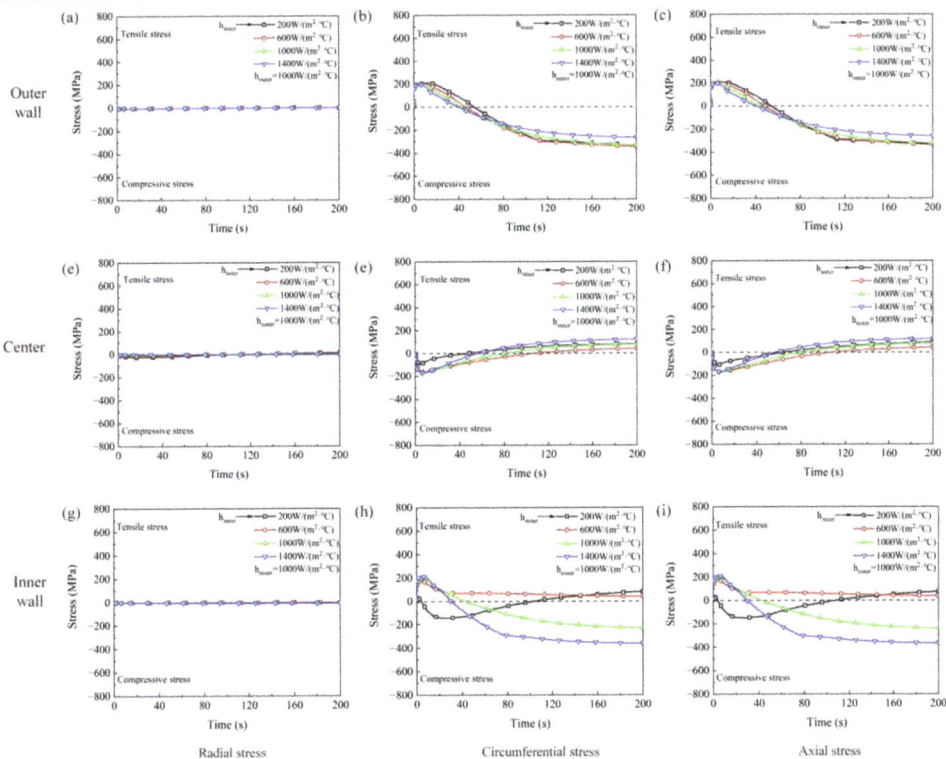

Figure 13. Stress-time curves of different positions with time at different cooling intensities of the inner wall. (**a**) radial stress of the outer wall; (**b**) circumferential stress of the outer wall; (**c**) axial stress of the outer wall; (**d**) radial stress of the center; (**e**) circumferential stress of the center; (**f**) axial stress of the center; (**g**) radial stress of the inner wall; (**h**) circumferential stress of the inner wall; (**i**) axial stress of the inner wall.

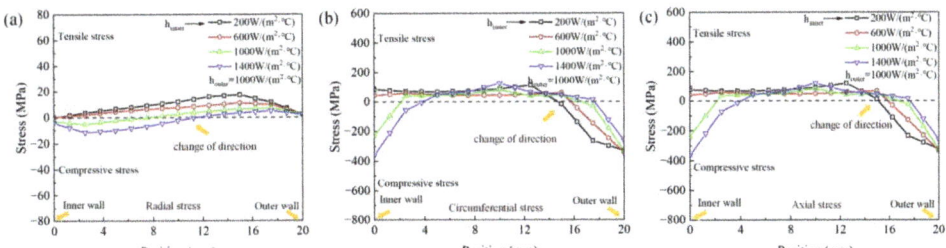

Figure 14. Stress distribution curve at different cooling intensities of the inner wall. (**a**) radial stress; (**b**) circumferential stress; (**c**) axial stress.

4. Conclusions

In this study, ANSYS simulation software was used to simulate the temperature and stress change of seamless steel tubes under different cooling intensities. By comparing the numerical simulation results of quenching residual stress of plate with the experimental data in reference [26], the accuracy of finite element simulation was verified. In addition, the temperature and stress curves of the steel tube were obtained by changing the cooling intensity of the inner and outer walls. The main conclusions are as follows:

With the increase in the cooling intensity of the outer wall, the cooling rate increases gradually. The cooling rate of the center increases firstly and then decreases, and gradually approaches the cooling rate of the inner wall.

During the quenching process, the direction of the stress component will change. The cooling intensity of the inner and outer walls directly determines the stress size and distribution. Therefore, by controlling the cooling intensity of the inner and outer walls, the stress distribution position and stress state of the steel tube section can be controlled.

When the cooling intensity of the inner wall is less than that of the outer wall, the radial stress is tensile stress, the circumferential and axial stress near the outer wall is compressive stress, and the circumferential and axial stress near the inner wall is tensile stress. When the cooling intensity of the inner wall is equal to or greater than that of the outer wall, the radial stress near the inner wall is compressive stress, and the radial stress near the outer wall is tensile stress. The circumferential and axial stresses near the inner and outer walls are compressive stresses, and the central position is tensile stress.

The residual stress increases with the increase in the cooling intensity of the outer wall. The maximum stress is mainly concentrated on the near wall of the steel tube. With the increase in the cooling intensity of the inner wall, the residual stress firstly decreases and then increases. When the inner and outer wall cooling intensity is the same, the residual stress obtains a minimum value.

Author Contributions: Conceptualization writing—review and editing, Z.L.; investigation, writing—original draft preparation, R.Z.; software, D.C.; methodology, Q.X.; validation, J.K.; project administration, funding acquisition, G.Y.; supervision, G.W. All authors have read and agreed to the published version of the manuscript.

Funding: This research was funded by the National Natural Science Foundation of China (No. U1860201, No.51804074).

Institutional Review Board Statement: Not applicable.

Informed Consent Statement: Not applicable.

Data Availability Statement: Not applicable.

Acknowledgments: Special thanks to National Natural Science Foundation of China, State Key Laboratory of Rolling and Automation, Northeastern University, for help and support.

Conflicts of Interest: The authors declare no conflict of interest.

References

1. Chen, H.T.; Yan, R. Development progress and reflections of seamless tube industry in China. *Steel Roll* **2014**, *31*, 41–44.
2. Wang, H.H.; Sang, W.; Chen, J.D.; Chen, D. Summary of Domestic Development of Seamless Steel Tube of for Large-capacity Gas Cylinder Service. *Steel Tube* **2019**, *47*, 7–13.
3. Yuan, G.; Kang, J.; Li, Z.L.; Wang, G.D. Development of Technology for Microstructure Control during On-line Heat Treatment Process for Hot-rolled Seamless Steel Tube. *Steel Tube* **2018**, *47*, 30–34.
4. Chen, Z.; Chen, X.; Zhou, T. Microstructure and Mechanical Properties of J55 ERW Steel Tube Processed by On-Line Spray Water Cooling. *Metals* **2017**, *7*, 150. [CrossRef]
5. Liu, Y.; Zuo, X.-w.; Chen, N.-l.; Rong, Y.-h.; Liu, S.-j. Finite element simulation of longitudinal quenching crack. *Trans. Mater. Heat Treat.* **2019**, *40*, 160–167.
6. Zhao, Y.-F.; Liu, H.-S. Analysis of thermal stress in inhomogeneous long cylindrical superconductor. *J. Lanzhou Univ. Technol.* **2018**, *44*, 167–172.
7. Souza, J.L.F.; Ziviani, M.; Vitor, J.F.A. Mathematical modeling of tube cooling in a continuous bed. *Appl. Therm. Eng.* **2015**, *89*, 80–89. [CrossRef]
8. Raič, J.; Landfahrer, M.; Klarner, J.; Zmek, T.; Hochenauer, C. Modelling of the cooling process of steel tubes in a rake type cooling bed. *Appl. Therm. Eng.* **2020**, *169*, 114895. [CrossRef]
9. Raič, J.; Landfahrer, M.; Klarner, J.; Zmek, T.; Hochenauer, C. Coupled CFD simulation for the cooling process of steel tubes in a rake type cooling bed. *Appl. Therm. Eng.* **2020**, *171*, 115068. [CrossRef]

10. Chen, D.; Zhang, R.; Li, Y.; Li, Z.; Yuan, G. Online Cooling System and Improved Similar Self-adaptive Strategy for Hot-rolled Seamless Steel Tube. *ISIJ Int.* **2021**, *61*, 2135–2142. [CrossRef]
11. Mascarenhas, N.; Mudawar, I. Methodology for predicting spray quenching of thick-walled metal alloy tubes. *Int. J. Heat Mass Transf.* **2012**, *55*, 2953–2964. [CrossRef]
12. Ali, M.A.; Hasan, S.T.; Myriounis, D.P. A Method of Analysis to Estimate Thermal Down-Shock Stress Profiles in Hollow Cylinders when Subjected to Transient Heat/Cooling Cycle. In *Advanced Materials Research*; Trans Tech Publications, Ltd.: Freienbach, Switzerland, 2012; pp. 893–898.
13. Schemmel, M.; Prevedel, P.; Schöngrundner, R.; Ecker, W.; Antretter, T. Size Effects in Residual Stress Formation during Quenching of Cylinders Made of Hot-Work Tool Steel. *Adv. Mater. Sci. Eng.* **2015**, *2015*, 678056. [CrossRef]
14. Hata, T.; Sumi, N. Thermal stress-focusing effect in a cylinder with phase transformation. *Acta Mech.* **2008**, *195*, 69–80. [CrossRef]
15. De Oliveira, W.P.; Savi, M.A.; Pacheco, P.M.C.L.; de Souza, L.F.G. Thermomechanical analysis of steel cylinders quenching using a constitutive model with diffusional and non-diffusional phase transformations. *Mech. Mater.* **2010**, *42*, 31–43. [CrossRef]
16. Chen, X.; Zhang, K.; Chen, G.; Luo, G. Multiple axial cracks in a coated hollow cylinder due to thermal shock. *Int. J. Solids Struct.* **2006**, *43*, 6424–6435. [CrossRef]
17. Devynck, S.; Denis, S.; Bellot, J.P.; Bénard, T.; Gradeck, M. Influence of boiling heat transfer and phase transformations on the deformation of a steel tube during quenching by impinging water jets. *Materialwiss. Werkst.* **2016**, *47*, 755–761. [CrossRef]
18. Leitner, S.; Winter, G.; Klarner, J.; Antretter, T.; Ecker, W. Model-based Residual Stress Design in Multiphase Seamless Steel Tubes. *Materials* **2020**, *13*, 439. [CrossRef]
19. Yang, X.W.; Zhu, J.C.; Lai, Z.H.; Yong, L.I.U.; Dong, H.E.; Nong, Z.S. Finite element analysis of quenching temperature field, residual stress and distortion in A357 aluminum alloy large complicated thin-wall workpieces. *Trans. Nonferr. Met. Soc. China* **2013**, *23*, 1751–1760. [CrossRef]
20. Brunbauer, S.; Winter, G.; Antretter, T.; Staron, P.; Ecker, W. Residual stress and microstructure evolution in steel tubes for different cooling conditions—Simulation and verification. *Mat. Sci. Eng. A Struct.* **2019**, *747*, 73–79. [CrossRef]
21. Gao, J.N.; Gao, Y.; Xu, Q.R.; Wang, G.; Li, Q. Simulation on flow, heat transfer and stress characteristics of large-diameter thick-walled gas cylinders in quenching process under different water spray volumes. *J. Cent. South Univ.* **2020**, *26*, 3188–3199. [CrossRef]
22. Xie, J.B.; He, T.C.; Cheng, H.M. Study on the Thermal Stress Field of Steel 1045 Quenched by Water. *Adv. Mater. Res.* **2011**, *255*, 3568–3572. [CrossRef]
23. Wang, X.; Li, F.; Yang, Q.; He, A. FEM analysis for residual stress prediction in hot rolled steel strip during the run-out table cooling. *Appl. Math. Model.* **2013**, *37*, 586–609. [CrossRef]
24. Şimşir, C.; Gür, C.H. 3D FEM simulation of steel quenching and investigation of the effect of asymmetric geometry on residual stress distribution. *J. Mater. Process. Technol.* **2008**, *207*, 211–221. [CrossRef]
25. Gradeck, M.; Kouachi, A.; Lebouché, M.; Volle, F.; Maillet, D.; Borean, J.L. Boiling curves in relation to quenching of a high temperature moving surface with liquid jet impingement. *Int. J. Heat Mass Transf.* **2009**, *52*, 1094–1104. [CrossRef]
26. Koc, M.; Culp, J.; Altan, T. Prediction of residual stresses in quenched aluminum blocks and their reduction through cold working processes. *J. Mater. Process. Technol.* **2006**, *174*, 342–354. [CrossRef]

Article

High-Cycle Fatigue Behavior and Fatigue Strength Prediction of Differently Heat-Treated 35CrMo Steels

Mengqi Yang [1,†], Chong Gao [2,3,†], Jianchao Pang [2,*], Shouxin Li [2], Dejiang Hu [1], Xiaowu Li [3] and Zhefeng Zhang [2,*]

1. Branch Company of Maintenance & Test, CSG Power Generation Co., Ltd., Guangzhou 511400, China; ymq9273@163.com (M.Y.); jianggehu@163.com (D.H.)
2. Shi-Changxu Innovation Center for Advanced Materials, Institute of Metal Research, Chinese Academy of Sciences, Shenyang 110016, China; ch_gao528@163.com (C.G.); shxli@imr.ac.cn (S.L.)
3. Key Laboratory for Anisotropy and Texture of Materials (Ministry of Education), Department of Material Physics and Chemistry, School of Materials Science and Engineering, Northeastern University, Shenyang 110819, China; xwli@mail.neu.edu.cn
* Correspondence: jcpang@imr.ac.cn (J.P.); zhfzhang@imr.ac.cn (Z.Z.)
† These authors contributed equally to this work.

Abstract: In order to obtain the optimum fatigue performance, 35CrMo steel was processed by different heat treatment procedures. The microstructure, tensile properties, fatigue properties, and fatigue cracking mechanisms were compared and analyzed. The results show that fatigue strength and yield strength slowly increase at first and then rapidly decrease with the increase of tempering temperature, and both reach the maximum values at a tempering temperature of 200 °C. The yield strength affects the ratio of crack initiation site, fatigue strength coefficient, and fatigue strength exponent to a certain extent. Based on Basquin equation and fatigue crack initiation mechanism, a fatigue strength prediction method for 35CrMo steel was established.

Keywords: 35CrMo steel; high-cycle fatigue; damage mechanism; fatigue strength prediction; heat treatment

1. Introduction

Chromium-molybdenum alloy steels (Cr-Mo steels) have been extensively applied in various industrial fields for their good mechanical properties, hydrogen resistance, and heat resistance. These fields include chemical industry, petrochemical industry, aviation industry, engineering vehicles, power industry, and many more [1,2]. The steels are mainly used to produce the parts of large equipment, such as safety valves, automobile clutches, pressure vessels [3], railway axles [4], gears [5,6], and bolts [7]. Most of these components are not only the independent parts of equipment, but are also subjected to cyclic loads. For instance, header bolts connect the engine's head cover with stay rings, and they are also subjected to pre-tightening loads and axial alternating loads from the head cover. Its reliability frequently determines the safe and stable operation of the engine subjected to complex loadings that can easily cause fatigue damage and may cause economic losses or even lead to major engineering accidents. In recent years, the fatigue research on Cr–Mo steels mainly focuses on the explorations of performance and mechanisms under extreme environments [3,8–12] or advanced technology [4,13,14]. However, there is little research on the prediction of fatigue strength for Cr–Mo steels. Therefore, the research on fatigue strength prediction of Cr–Mo steels cannot be ignored.

In addition, Cr-Mo steels can also be machined into components with different performance requirements, e.g., wear-resistant components with high hardness and high strength [15], mill liners with wear properties and impact toughness [16], shock-resisting tools with the superior combination of hardness and impact properties [17], bolts with high comprehensive mechanical properties [18], etc. Heat treatment is the main technique to achieve these properties by regulating the microstructures or surface chemical composition. For example, quenching can improve the hardness and wear resistance of steel; and the different tempering temperatures can obtain different strengths and toughness [19,20]. Therefore, in the process of designing heat treatment procedures of materials for the components, it is necessary to adjust and test their mechanical properties and fatigue performance. However, fatigue test is time and energy consuming, so it is important to predict fatigue strength from static mechanical properties.

The main methods of fatigue strength prediction are El Haddad et al.'s model and Murakami's $\sqrt{area}$ parameter model [21]. However, they have certain limitations, the former has no estimation method for 3-D inclusions; the latter believes that the same material has defects with the same size, and it has no effective estimate for internal and unknown size defects. Therefore, it is still necessary to explore the fatigue strength prediction method for engineering materials from the fatigue curve (S–N curve).

In the early 20th century, researchers found the linear relation between stress amplitude and life on log–log plots, and proposed a simple formula such that

$$\sigma_a = \sigma_f'(2N_f)^b \qquad (1)$$

where σ_a is the stress amplitude, σ_f' is the fatigue strength coefficient, b is the fatigue strength exponent, and N_f is the number of cycles to failure. The values of fatigue strength coefficient and fatigue strength exponent are the intercept and slope of the S–N curves, respectively, on log–log plots. Nowadays, this is the well-known Basquin equation, and it has become an important tool for determining the fatigue strength and design criterion of materials. In recent years, the characteristics of the S–N curve and Basquin equation have been studied by many investigators [22–24]. Some researchers have proposed formulas to estimate the values of σ_f' and b, which are generally based on the inclusion size, hardness, and tensile strength [25–27]. However, the shape of S–N curve and the values of σ_f' and b could be changed by many other factors, such as sample surface treatment, experimental environment, and loading type [22,28,29]. It is valuable to further explore the high cycle fatigue (HCF) strength prediction of Cr–Mo steels.

In this study, four heat-treatment procedures of 35CrMo (Chinese designation) steel were employed to investigate the microstructures, tensile and HCF behaviors, and the relations among them. The differences in the mechanical behaviors of variously heat-treated 35CrMo steels were also analyzed. According to the corresponding fracture mechanisms, a suitable formula of fatigue strength prediction for the Cr–Mo steel was established.

2. Experimental Materials and Procedures

The chemical composition of 35CrMo steel is shown in Table 1. To gain a wide range of strength, the as-received steel bars were heated at 860 °C for 30 min followed by the oil-quenching. Then, some of the steel bars were processed into specimens, and the rest of them were tempered at 200 °C, 400 °C, and 500 °C for 90 min, respectively, followed by air-cooling to room temperature. The four heat-treatment procedures are given in Table 2, and the corresponding specimens are named as Q, QT200, QT400, and QT500, respectively.

Table 1. Chemical composition of 35CrMo/%.

C	Si	Mn	Cr	Mo	P	S	Fe
0.35	0.35	0.76	1.13	0.20	<0.005	<0.001	Balance

The dimensions of the tensile and fatigue specimens are shown in Figure 1. Tensile tests were conducted at a strain rate of 10^{-3} s^{-1} by an Instron 5982 static testing machine (Instron Corporation, Boston, MA, USA). The HCF tests were conducted under symmetrical push-pull loading condition (R = −1) by using a GPS100 high-frequency fatigue tester

(Sinotest Equipment Co., Ltd., Changchun, China) under room temperature in air. The HCF tests were proceeded at a resonance frequency of about 115 Hz. In this experiment, about 20 specimens were prepared for each heat-treatment condition. Tests were stopped when the specimen failed completely or achieved 10^7 cycles. The fatigue strength was determined using the staircase method in which five pairs of specimens were tested, namely, taking the average values of these stress levels. The S–N curves were fitted with the data of all failed specimens by the least square method, which means that half of the specimens could fail above the curves [30]. The fatigue strength coefficients and exponents were obtained by the same method.

Table 2. Heat-treatment procedures of 35CrMo steel.

Samples	Quenching	Tempering
Q		Untempered
QT200	Preheating to 860 °C for 30 min and	200 °C tempering for 90 min
QT400	quenching in oil	400 °C tempering for 90 min
QT500		500 °C tempering for 90 min

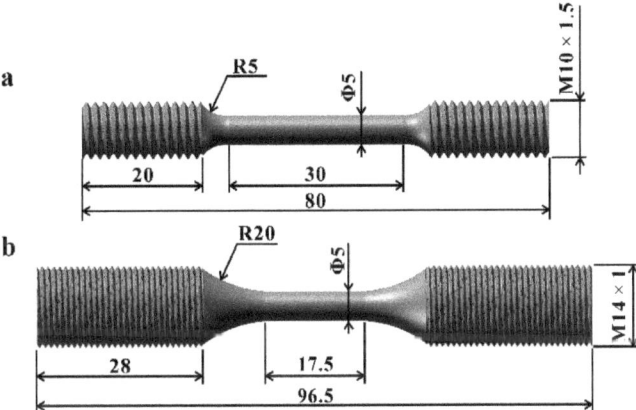

Figure 1. Configurations and dimensions of specimens tested for tensile (a) and fatigue (b) properties. (Unit: mm).

The microstructures of specimens with different heat-treatment procedures were examined by electron back scattered diffraction (EBSD, LEO Supra 35, Carl Zeiss AG, Oberkochen, Germany). The tensile and fatigue fracture surfaces of failed specimens were examined by scanning electron microscopy (SEM, JSM-6510, Japan Electronics Co., Ltd., Tokyo, Japan).

3. Results and Discussion

3.1. Microstructure

The EBSD microstructures of 35CrMo steel with four heat-treatment procedures are shown in Figure 2. It can be seen that Q specimen contains many lath martensites and some retained austenites. The microstructure of QT200 specimen consists of plate shaped tempered martensites and some retained austenites. Both QT400 and QT500 specimens display the uniform microstructures of tempered troostite, as shown in Figure 2c,d.

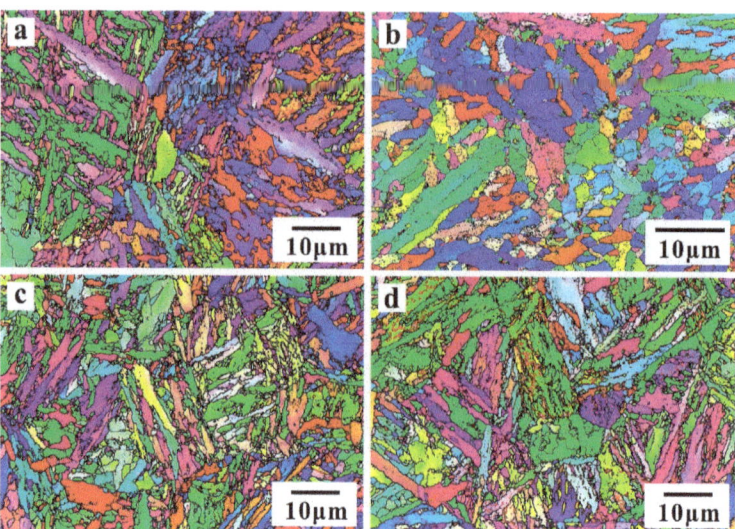

Figure 2. EBSD microstructures for 35CrMo steel with four heat-treatment procedures. (a) Q, (b) QT200, (c) QT400, and (d) QT500.

3.2. Tensile Behaviors

The tensile properties of 35CrMo with different heat-treatment procedures are provided in Figure 3. The tensile properties of 35CrMo steel at different tempering temperatures are listed in Table 3. As can be seen from Figure 3b, with the tempering temperature increasing, the tensile strength (σ_b) successively decreases; besides, the yield strength (σ_y) slowly increases at first and then decreases, which are in agreement with the cases of other steels [19,31]. It is observed that the percentage reduction of area (Z) and elongation after fracture (A) increase in different degrees with increasing tempering temperature as shown in Figure 3c. Figure 3d gives relations of the elongation after fracture and the percentage reduction of area versus the tensile strength of 35CrMo steel. As the tensile strength increases, the elongation after fracture and the percentage reduction of area decrease in varying degrees. This is consistent with the inverse relation between strength and ductility for lots of metals [19].

Table 3. Tensile properties for 35CrMo steel processed at different tempering temperatures.

Sample	σ_b/MPa	σ_y/MPa	Z/%	A/%
Q	1977	1380	33.20	10.80
QT200	1891	1487	47.66	12.05
QT400	1566	1352	51.84	12.10
QT500	1261	1170	58.53	16.20

The macroscopic fractographies of tensile specimens for 35CrMo steel are shown in Figure 4. It can be seen that the tensile specimens with different tempering temperatures have significant necking phenomena. With the increase of tempering temperature, the area ratio of fiber zone (the ratio of the fiber zone area to the fracture surface area) gradually decreases, and the area ratio of shear lip first increases and then decreases slightly. Q and QT200 specimens have no obvious radial pattern, as shown in Figure 4a,b. QT400 and QT500 specimens have radial zone, the area ratio of radial zone increases and radial pattern becomes pronounced with the increase of tempering temperature, as shown in Figure 4c,d.

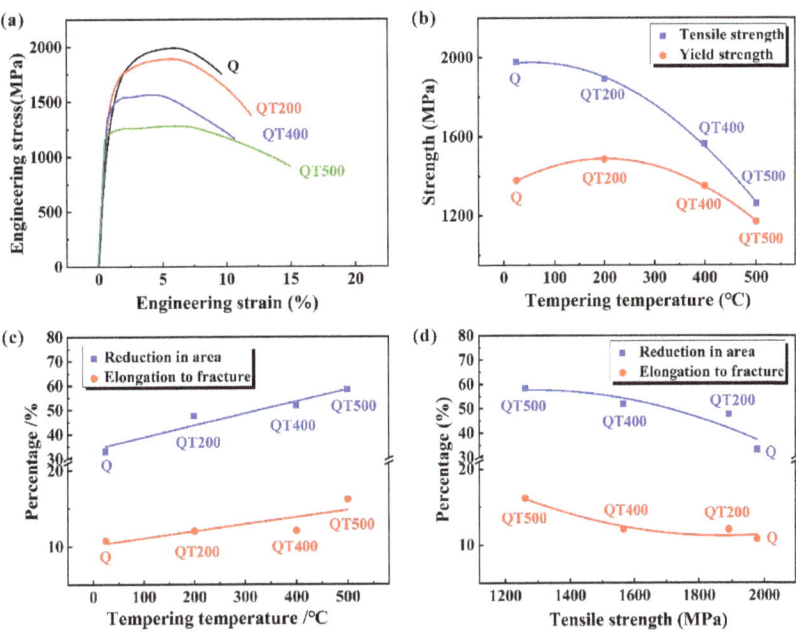

Figure 3. Tensile properties of 35CrMo steel. (**a**) Tensile engineering stress-strain curves; (**b**,**c**) the relation between strengths (tensile and yield strengths), percentages (elongation after fracture and percentage reduction of area), and tempering temperature; and (**d**) relations of percentages vs. tensile strength.

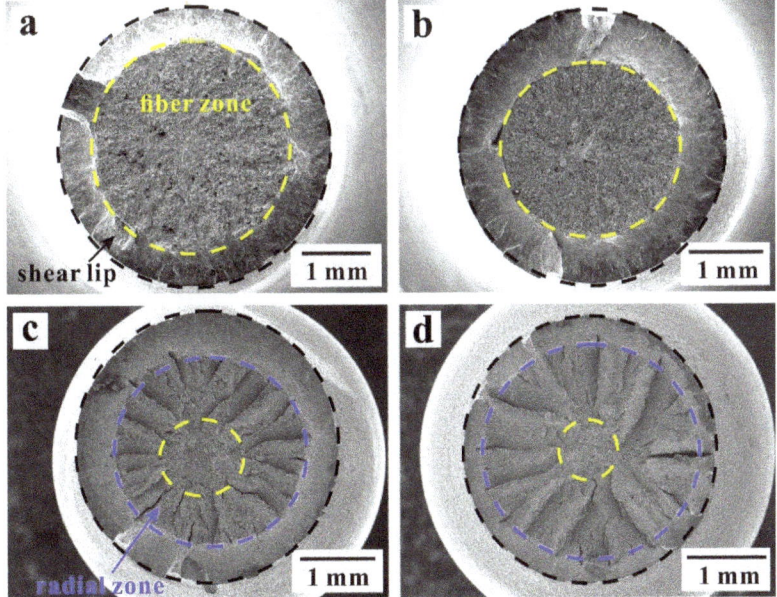

Figure 4. The macroscopic fractographies of tensile samples for 35CrMo steel processed at different tempering temperatures. (**a**) Untampered, (**b**) 200 °C, (**c**) 400 °C, and (**d**) 500 °C.

Tensile fractographies in the fiber zone for 35CrMo steel are magnified in Figure 5. It can be seen that the fiber zones of these specimens are mainly composed of dimples with different sizes, implying the typical ductile fracture modes. Besides, few microcracks and some larger voids can also be seen from the figure. The formation of microcracks and voids in the fiber zone can be attributed to the transition of the stress states of the specimen from uniaxial to triaxial due to the necking of specimens. The plastic deformation at the axial center of the specimen is difficult to continue with the effect of triaxial stress, so that the stress concentration occurs at the inclusions or second-phase particles, where the voids eventually nucleate and grow. Consequently, the sizes of microcracks or voids are closely related to inclusions or second-phase particles. It can be noted from Figure 5 that the sizes of microcracks and voids increase with the increase of tempering temperature, and such a similar situation has also appeared in high-strength, high ductility steels [32]. It can be concluded that the strength and toughness affect the behaviors of inclusions or second-phase particles. This seems to be consistent with the effect of tensile loads on the behaviors of inclusion and second-phase particles at elevated temperature, which is due to the transformation of tensile properties affected by high temperature [33].

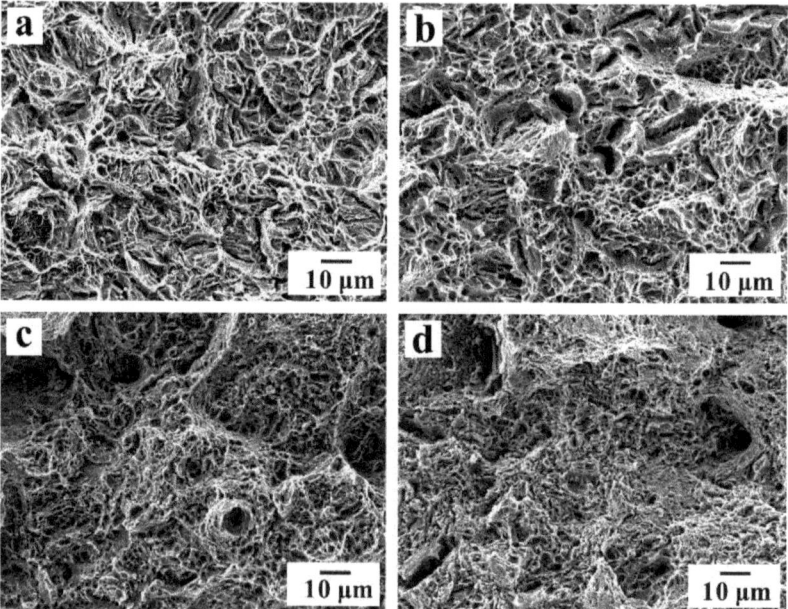

Figure 5. Tensile fractographies in the fiber zone for 35CrMo steel processed at different tempering temperatures. (**a**) Untampered, (**b**) 200 °C, (**c**) 400 °C, and (**d**) 500 °C.

3.3. High-Cycle Fatigue Behaviors

The S–N curves of 35CrMo steel under different heat treatments are shown in Figure 6a. The fatigue properties of 35CrMo steel at different tempering temperatures are listed in Table 4. Obviously, QT200 specimens have the best fatigue resistance. The fatigue strengths (σ_w) increase first and then decrease with the increase of tensile strengths (Figure 6b), which were also found in many other materials [19,34]. The Basquin equations for these materials are as below (Equations (2)–(5)):

$$\sigma_a = 2040.42(2N_f)^{-0.073}, \text{ for Q} \tag{2}$$

$$\sigma_a = 1718.57(2N_f)^{-0.058}, \text{ for QT200} \tag{3}$$

$$\sigma_a = 2261.03(2N_f)^{-0.089}, \text{ for QT400} \qquad (4)$$

$$\sigma_a = 2539.02(2N_f)^{-0.126}, \text{ for QT500} \qquad (5)$$

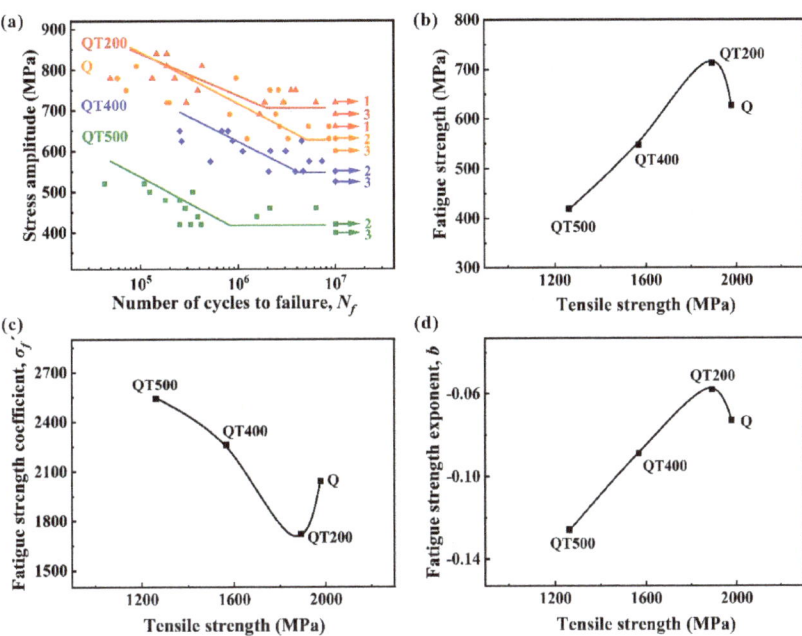

Figure 6. (**a**) Relations between stress amplitude and fatigue life (S–N curves), (**b**) the relation between tensile and fatigue strengths, (**c**) the relation between fatigue strength coefficient and tensile strength, and (**d**) the relation between fatigue strength exponent and tensile strength.

Table 4. Fatigue properties and S–N curves parameters for 35CrMo steel processed at different tempering temperatures.

Sample	σ_w/MPa	σ_f'	b
Q	627	2040.42	−0.073
QT200	706	1718.57	−0.058
QT400	548	2261.03	−0.089
QT500	418	2539.02	−0.126

In Equations (2)–(5), the obtained fatigue strength coefficient σ_f' and fatigue strength exponents b are reported for the considered cases. The relations of fatigue parameters (σ_f' and b) vs. the tensile strengths are shown in Figure 6c,d. It can be seen that the increasing and decreasing trends of them are opposite and both curves have extreme values at data of QT200 specimens. This is inconsistent with the trend of steels for very high cycle fatigue (VHCF) [27]. Some researchers pointed out that HCF and VHCF behaviors are different for the same materials [22,35,36]. Therefore, it is essential to study the variations of the fatigue strength coefficient and exponent in a wide strength range from the perspective of HCF.

The fatigue strength coefficient and the fatigue strength exponent are mainly affected by strengthening mechanisms and damage mechanisms of materials respectively [27]. In order to understand the variation trends of fatigue strength coefficient and exponent for 35CrMo steel, it is necessary to study the fracture mechanism of failed specimens. The fatigue source regions of failed specimens with different heat-treatment procedures were observed by SEM. According to different crack initiation mechanisms, these specimens could be divided into five categories, as shown in Figure 7, such as (a) surface scratch; (b) surface inclusion; (c) subsurface inclusion, representing the inclusion whose distance from the surface is less than its own size in this paper; (d) inner inclusion, representing the inclusion whose distance from the surface is greater than its size; and (e) micro-facet comprising numerous small convex and concave, representing the trace of plastic deformation caused by non-inclusion crack [37,38]. For the convenience of statistics, some researchers have summarized the fatigue crack initiation sites into two types, namely, surface and inner [19]. Inner represents inner inclusion and micro-facet, and surface scratch, surface inclusion, and subsurface inclusion are classified as surface, as shown in Figure 8.

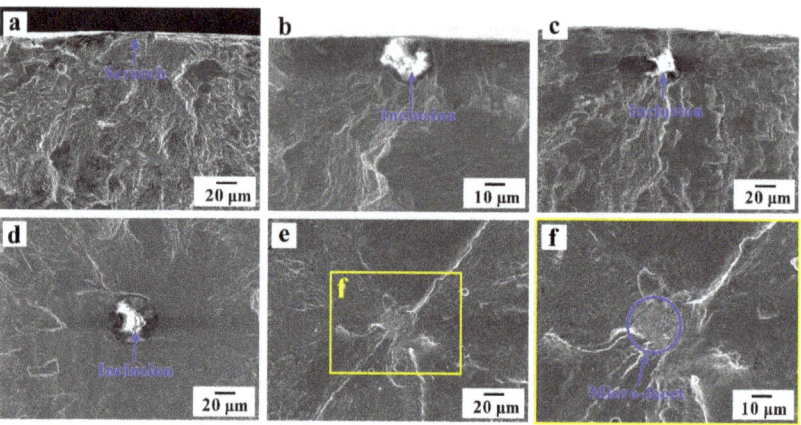

Figure 7. Fatigue crack initiation morphologies. (**a**) Surface scratch, (**b**) surface inclusion, (**c**) subsurface inclusion, (**d**) inner inclusion, and (**e**,**f**) micro-facet.

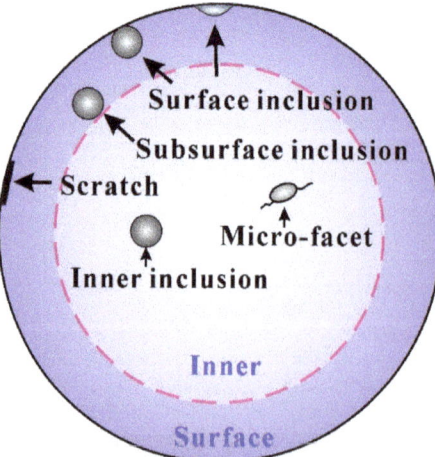

Figure 8. Schematic diagram of crack initiation site.

The two types of failed specimens have been indicated in the S–N curves, as shown in Figure 9. In the figure, the circles represent the failed specimens with cracks initiated on the surface and the solid circles represent the cracks initiated inside. It is found that the specimens with initiation of inner cracks are generally loaded at low-stress levels and have high fatigue life, which can be clearly seen in Figure 9a,b. The same situation has also been found by some other researchers [13,22,37]. Under high applied stress amplitude, the surface defects and processing defects are the obvious weak zones, since the plastic deformation preferentially occurs at surface due to lack of constrain. The locally accumulated plastic strain caused by high stress concentration at the surface defects and processing defect will induce crack initiation. On the other hand, when the lower stress amplitude is applied, the locally accumulated plastic strain over those surface defects becomes weaker; at this time, some interior inclusions may have the potential to compete with those defects. Since the inner area of a cross section is generally much larger than the outer surface layer area, the probability for larger inclusions or harmful inclusions emerging in the inner area is definitely greater than that in the surface area. If so, the fatigue cracks may initiate from internal inclusions at the low stress amplitude.

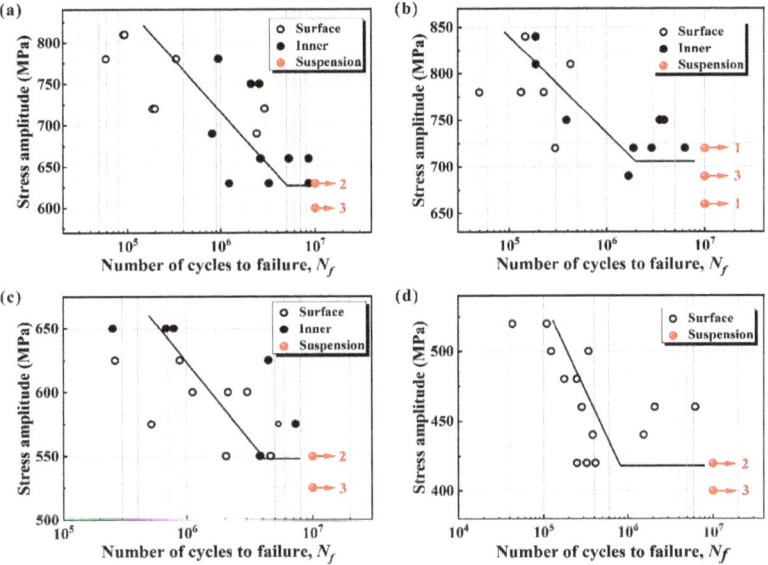

Figure 9. S–N curves for the specimens of Q (a), QT200 (b), QT400 (c) and QT500 (d).

From Figure 9, one can see that most of the failure samples for Q begin to fracture from the inside, and the number of such failed samples gradually decreases with the increase of tempering temperature of heat-treatment procedures. Until the tempering temperature reaches 500 °C, all the failed specimens begin to fracture on the surface. Figure 10a shows the relations between the ratios of surface/inner fatigue crack initiation sites (the ratios of the number of failures originating from the surface/inner to the total number of failures) and yield strengths. It can be seen that the ratio of surface initiation cracks decreases with the increase of yield strength. In other words, as the yield strength decreases, the trend of surface fatigue crack initiation increases. It is understood with lower yield strength, the severe locally accumulated plastic deformation will easily result in the surface defects as mentioned above. Furthermore, it can be roughly inferred from the figure that cracks will initiate from the surface for the specimens with yield strengths below 1200 MPa. The ratio of inner cracks will continue to increase when the yield strengths of the samples are higher than 1500 MPa. To sum up, it can be said that the yield strength affects the ratio of fatigue crack initiation site to a certain extent.

Wang et al. [39] have concluded that the transition from surface to subsurface crack initiation has a significant effect on the slope of S–N curve. As an extension, the intercepts and slopes of S–N curves (fatigue strength coefficient and exponent of Basquin equation) are related to fatigue crack initiation sites, as shown in Figure 10b. It can be seen that the fatigue strength coefficient decreases and the fatigue strength exponent increases with the increasing ratio of the inner crack site. Therefore, different ratios of crack initiation sites affect the fatigue strength coefficient and exponent of Basquin equation to a certain extent. The reason can be found from the distribution characteristics of different crack initiations sites in Figure 9 and the relations in Figure 10b. Combined with the above conclusions that the yield strength affects the ratio of fatigue crack initiation site and the cracking position affects the fatigue strength coefficient and exponent, it can be said that the fatigue strength coefficient and exponent are indirectly influenced by the yield strength.

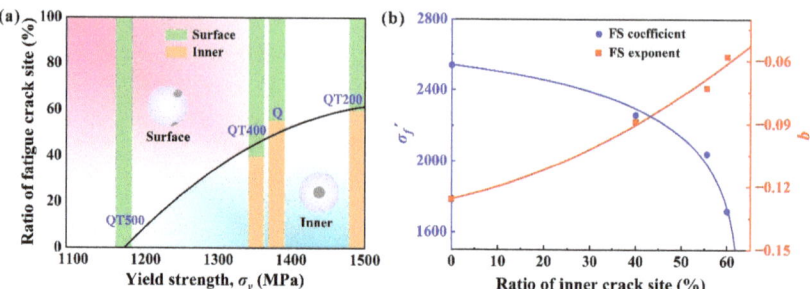

Figure 10. (a) The relation between the ratios of fatigue crack initiation sites and yield strength, and (b) relations of the fatigue strength coefficient and exponent vs. ratio of inner crack site.

3.4. Prediction of Fatigue Strength

To predict fatigue strength by Basquin equation, some parameters are necessary to figure out. As shown in Figure 11a, the fatigue strength σ_w of a material can be determined by the fatigue strength coefficient, exponent, and the life of knee point N_k in the S–N curve. The knee point is the intersection of the curve fitted by the group method and the fatigue strength calculated by the staircase method. Obviously, the knee point is also a necessary parameter to predict fatigue strength.

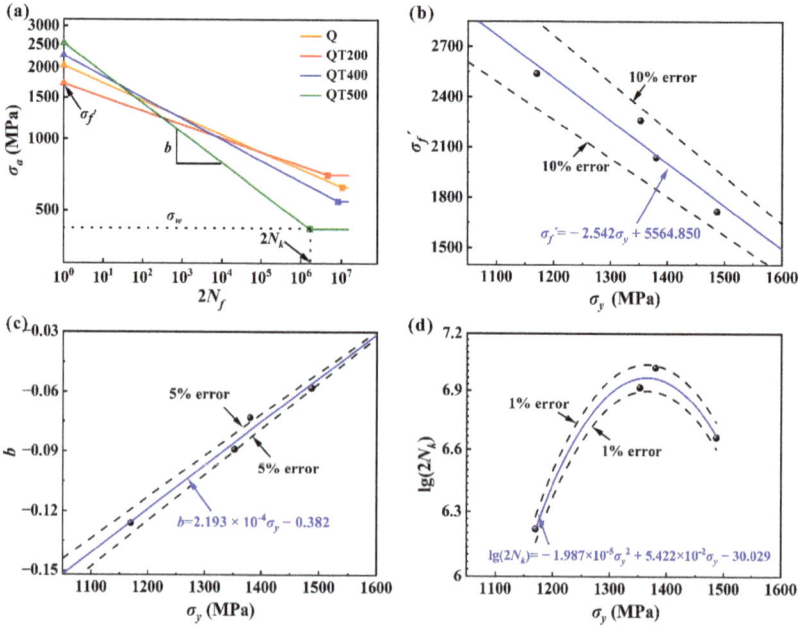

Figure 11. The fatigue strength prediction model. (a) The schematic illustration of S–N curves, (b) the linear relation between σ_f' and σ_y, (c) the linear relation between b and σ_y, and (d) the relation of $\lg(2N_k)$ and σ_y.

The logarithmic form of Basquin equation for S–N curves can be obtained,

$$\lg \sigma_a = b \lg(2N_f) + \lg \sigma'_f \tag{6}$$

If N_k is determined, the fatigue strength prediction equation can be written as

$$\lg \sigma_w = b \lg(2N_k) + \lg \sigma'_f \tag{7}$$

Based on the above discussion, σ'_f and b are linearly fitted with the yield strength, and the error bands are within the 10% and 5%, respectively, as shown in Figure 11b,c. In addition, the knee point is also fitted with the yield strength for the unification of variables and convenience of calculation. They have a quadratic relation with only 1% error band, as shown in Figure 11d. This is the relation between the intersection of the two lines and the yield strength, which has no practical significance. The fitting equations can be expressed in linear and quadratic equations as below, respectively,

$$\sigma'_f = m\sigma_y + n \tag{8}$$

$$b = u\sigma_y + v \tag{9}$$

$$\lg(2N_k) = x\sigma_y^2 + y\sigma_y + z \tag{10}$$

Substituting Equations (8)–(10) into Equation (7), a new relation can be obtained,

$$\lg \sigma_w = (u\sigma_y + v)(x\sigma_y^2 + y\sigma_y + z) + \lg(m\sigma_y + n) \tag{11}$$

where, $m, n, u, v, x, y,$ and z are the material constants, which can be obtained by data fitting.

For 35CrMo steel, the constants have been fitted and the fatigue strength prediction formula can be expressed as follows,

$$\lg \sigma_w = (2.193 \times 10^{-4}\sigma_y - 0.382)(-1.987 \times 10^{-5}\sigma_y^2 + 5.422 \times 10^{-2}\sigma_y - 30.029) + \lg(-2.542\sigma_y + 5564.850) \tag{12}$$

The results of fatigue strength prediction are shown in Figure 12, and it can be seen that the errors of this fatigue prediction equation are less than 10%.

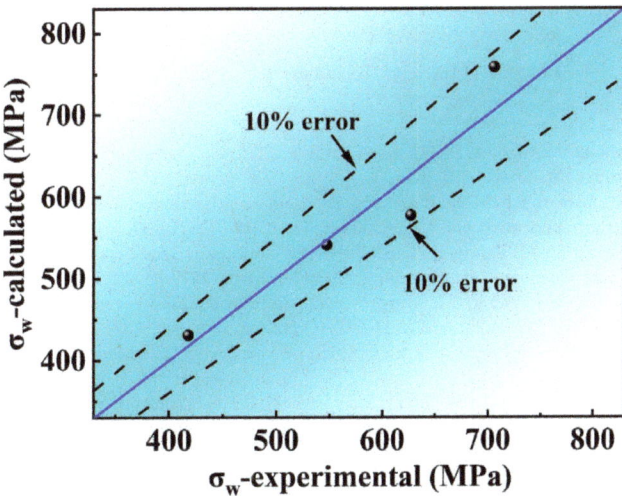

Figure 12. The calculated vs. experimental values for fatigue strength.

4. Conclusions

The fatigue fracture morphologies and HCF properties of 35CrMo steel specimens with different tensile strengths were studied. The main conclusions can be summarized as below:

(1) With the increase of tempering temperature, martensite is gradually decomposed and the tensile strength decreases, but the yield strength and fatigue strength increase at first and then decrease. QT200 specimens have the best fatigue performance;
(2) To some extent, the yield strength affects the ratio of crack initiation site for a specimen, and the crack initiation site affects the fatigue strength coefficient and fatigue strength exponent. Therefore, the yield strength affects the change of fatigue strength coefficient and fatigue strength exponent, and they have a linear relation for HCF tests of 35CrMo steel;
(3) A fatigue strength prediction method based on the damage mechanisms and Basquin equation was proposed. In this way, the values of fatigue strength coefficient, fatigue strength exponent, and knee point can be expressed by yield strength. This method can effectively predict the HCF strength of 35CrMo steel. The fatigue strength coefficient, fatigue strength exponent, and knee point are affected by many factors, and it is still necessary to further explore whether this method is suitable for other materials.

Author Contributions: Conceptualization, J.P. and Z.Z.; Funding acquisition, M.Y., J.P., D.H. and Z.Z.; Investigation, C.G.; Methodology, C.G., S.L., J.P., and X.L.; Project administration, M.Y., and D.H.; Resources, M.Y., and D.H.; Writing-original draft, C.G.; Writing—review & editing, J.P., S.L., X.L., and Z.Z. All authors have read and agreed to the published version of the manuscript.

Funding: This research was funded by project entitled "Research on Lifetime Prediction of Non-rotating Parts of Pump Turbine Unit Based on Rotor-Stator Interaction (RSI), Fluid-Structure Coupling and Fracture Mechanics" under Grant. No. 022200KK52180006, National Natural Science Foundation of China (NSFC) under Grant. No. 51871224, Natural Science Foundation of Liaoning Province under Grant. No. 20180550880.

Institutional Review Board Statement: The study did not require ethical approval.

Informed Consent Statement: Not applicable.

Data Availability Statement: Not applicable.

Acknowledgments: The authors would like to thank Z. K. Zhao and H. Y. Zhang for their help of the fatigue experiment and SEM observations.

Conflicts of Interest: The authors declare no conflict of interest.

References

1. Tanaka, K.; Shimonishi, D.; Nakagawa, D.; Ijiri, M.; Yoshimura, T. Stress relaxation behavior of cavitation-processed Cr-Mo steel and Ni-Cr-Mo steel. *Appl. Sci.* **2019**, *9*, 299. [CrossRef]
2. Raj, B.; Choudhary, B.; Raman, R.S. Mechanical properties and non-destructive evaluation of chromium–molybdenum ferritic steels for steam generator application. *Int. J. Press. Vessel. Pip.* **2004**, *81*, 521–534. [CrossRef]
3. Ma, K.; Zheng, J.; Hua, Z.; Gu, C.; Zhang, R.; Liu, Y. Hydrogen assisted fatigue life of Cr–Mo steel pressure vessel with coplanar cracks based on fatigue crack growth analysis. *Int. J. Hydrogen Energy* **2020**, *45*, 20132–20141. [CrossRef]
4. Zhang, J.; Lu, L.; Wu, P.; Ma, J.; Wang, G.; Zhang, W. Inclusion size evaluation and fatigue strength analysis of 35CrMo alloy railway axle steel. *Mater. Sci. Eng. A* **2013**, *562*, 211–217. [CrossRef]
5. Lv, Y. Influence of laser surface melting on the micropitting performance of 35CrMo structural steel gears. *Mater. Sci. Eng. A* **2013**, *564*, 1–7. [CrossRef]
6. Takemasu, T.; Koide, T.; Shinbutsu, T.; Sasaki, H.; Takeda, Y.; Nishida, S. Effect of Surface Rolling on Load Bearing Capacity of Pre-alloyed Sintered Steel Gears with Different Densities. *Procedia Eng.* **2014**, *81*, 334–339. [CrossRef]
7. Shi, H.Q.; Ding, Y.; Ma, L.Q.; Shen, X.D. Corrosion Failure Analysis of 35CrMo Bolt in Wet Hydrogen Sulfide Environment. *Appl. Mech. Mater.* **2013**, *291–294*, 2605–2609. [CrossRef]
8. Gaur, V.; Doquet, V.; Persent, E.; Roguet, E. Effect of biaxial cyclic tension on the fatigue life and damage mechanisms of Cr–Mo steel. *Int. J. Fatigue* **2016**, *87*, 124–131. [CrossRef]

strength was negligible. Luke et al. [19] made conclusions on some important aspects and results related to the application of the fracture mechanics approach to the prediction of inspection intervals of railway axles under in-service conditions.

The specimen geometry is an important factor affecting fatigue properties [20–22]. The fatigue performance tends to decrease with the increase of specimen size [23,24]. The size and shape of actual components are usually different from the standard testing specimens. Therefore, studying the size and shape effects on the fatigue behavior of materials in key structures of high-speed trains has scientific significance and application value. Li et al. [25] studied the effects of specimen size and notch on the fatigue properties of an EA4T axle steel. The study indicated that, with the increase of specimen size, the fatigue strength of the dogbone specimen was considerably lower than that of the hourglass specimen under axial loading. Shen et al. [26] analyzed the effect of inclusion size on the fatigue strength of small specimens and railway axles, and showed that, due to the increase of risk volume, the critical stress of fatigue failure in axles induced by inclusion was about 50% of that in small specimens under rotating bending loading. Varfolomeev et al. [27] studied the effect of specimen shape on the fatigue crack growth rate of an EA4T axle steel and showed that the crack growth rate depended on the specimen shape and loading condition.

Axle box bodies are important components in high-speed trains, which are subject to cyclic loadings and might fail in service. However, there are few results available for the effects of specimen size and shape on the fatigue behavior of materials for axle box bodies. Therefore, revealing the size and shape effects on the fatigue behavior of materials for axle box bodies is of great importance. This paper studies G20Mn5QT steel from axle box bodies in high-speed trains. The axial loading fatigue tests are at first conducted on the specimens with different size and shape. Then, the fatigue failure mechanism of G20Mn5QT steel is studied based on the observation of the fracture surface by scanning electron microscope (SEM) and the analysis of the crack initiation region by energy dispersive X-ray spectroscopy (EDS). Finally, the size and shape effects on the fatigue performance are correlated by using the probabilistic control volume method for G20Mn5QT steel.

2. Materials and Methods

The material used is a G20Mn5QT steel cut from the new axle box bodies of a high-speed train. The chemical composition is 0.18 C, 0.34 Si, 1.20 Mn, 0.22 Ni, 0.065 Al, 0.03 Cr, 0.011 Cu, 0.017 P, and 0.009 S in weight percent (Fe balance). The axle box body was at first heated at $910 \pm 10°$ for 3.5 h and oil quenched, and then it was tempered for 4 h at $640 \pm 10°$ and cooled to below 300 °C with the furnace and then air cooled. The average tensile strength and yield strength of the material are 582 MPa and 399 MPa, respectively. The standard deviation is 0.58 for the tensile strength and 5.5 for the yield strength. They are obtained from three specimens by an MTS Landmark machine. The strain rate is 5×10^{-4} s^{-1}. The shape of the tension specimen is shown in Figure 1a. Fatigue tests were conducted on an MTS Landmark machine. The loading frequency is 1 Hz to 32 Hz and the stress ratio R is -1. Two kinds of specimens, the hourglass specimen and the dogbone specimen, are chosen for fatigue tests, as shown in Figure 1b,c, respectively. The elastic stress concentration factor K_t is defined as the ratio of the maximum principal stress at the notch root to that of the cylindrical specimen with the same smallest cross section (i.e., nominal stress), and is obtained by using Abaqus 6.14 software. In the calculation, Young's modulus is $E = 210$ GPa and Poisson's ratio is $\nu = 0.3$. All tests were carried out at room temperature in air. Before the fatigue test, the surface of the experimental section of the specimen was ground and polished. The surface roughness R_a was less than 0.5 μm.

SEM is used to observe the fatigue fracture surface and analyze the crack initiation mechanism. EDS is conducted to determine the element composition in the typical crack initiation region.

Article

Size and Shape Effects on Fatigue Behavior of G20Mn5QT Steel from Axle Box Bodies in High-Speed Trains

Zhenxian Zhang [1], Zhongwen Li [1], Han Wu [2,3] and Chengqi Sun [2,3,*]

[1] CRRC Qingdao Sifang Co., Ltd., Qingdao 266111, China; zhangzhenxian@cqsf.com (Z.Z.); lzhw1980@163.com (Z.L.)
[2] State Key Laboratory of Nonlinear Mechanics, Institute of Mechanics, Chinese Academy of Sciences, Beijing 100190, China; wuhan@lnm.imech.ac.cn
[3] School of Engineering Science, University of Chinese Academy of Sciences, Beijing 100049, China
* Correspondence: scq@lnm.imech.ac.cn; Tel.: +86-10-8254-3968

Abstract: In this paper, the axial loading fatigue tests are at first conducted on specimens of G20Mn5QT steel from axle box bodies in high-speed trains. Then, the size and shape effects on fatigue behavior are investigated. It is shown that the specimen size and shape have an influence on the fatigue performance of G20Mn5QT steel. The fatigue strength of the hourglass specimen is higher than that of the dogbone specimen due to its relatively smaller highly stressed region. Scanning electron microscope observation of the fracture surface and energy dispersive X-ray spectroscopy indicate that the specimen size and shape have no influence on the fatigue crack initiation mechanism. Fatigue cracks initiate from the surface or subsurface of the specimen, and some fracture surfaces present the characteristic of multi-site crack initiation. Most of the fatigue cracks initiate from the pore defects and alumina inclusions in the casting process, in which the pore defects are the main crack origins. The results also indicate that the probabilistic control volume method could be used for correlating the effects of specimen size and shape o the fatigue performance of G20Mn5QT steel for axle box bodies in high-speed trains.

Keywords: G20Mn5QT steel; crack initiation mechanism; fatigue strength; size effect; shape effect

1. Introduction

The high-speed railway industry has developed rapidly in the past decade. Fatigue failure, as one of the main failure modes for engineering materials and components [1–5], is also a key mechanical problem for high-speed trains. Many studies concerning the fatigue problems in high-speed trains have been carried out [6–11]. For example, Lu et al. [12] studied the very-high-cycle fatigue behavior of an axle steel LZ50 under rotating bending fatigue loading and showed that LZ50 steel had the fatigue limit at $5 \times 10^6 \sim 10^9$ cycles. The fatigue fracture surface observation indicated that the fatigue crack initiated from the ferrite on the surface of the specimen. Chen et al. [13] investigated the high cycle and very-high-cycle fatigue performance of an axle steel EA4T, and found that there was still a conventional fatigue limit for EA4T steel. Beretta et al. [14] studied the corrosion fatigue behavior of an axle steel A1N exposed in rainwater, and the results showed that the rainwater significantly reduced the fatigue strength (>10^6 cycles) of the A1N steel. Wang et al. [15] analyzed the fatigue strength of the CRH2 motor bogie frame through simulation and online tests. Zhang et al. [16] studied the fatigue crack growth behavior in the gradient microstructure of the surface layer of S38C axle steel. The results indicated that the crack growth rate firstly decelerated and then accelerated with increasing the crack length in the gradient layer. Guagliano and Vergani [17] conducted experiments and numerical analysis on the sub-surface cracks in railway wheels. Gao et al. [18] studied the effect of artificial defects on the fatigue strength of an induction hardened S38C axle and showed that the influence of shallower impact damage (smaller than 200 μm) on fatigue

38. Gui, X.L.; Gao, G.H.; An, B.F.; Misra, R.D.K.; Bai, B.Z. Relationship between non-inclusion induced crack initiation and microstructure on fatigue behavior of bainite/martensite steel in high cycle fatigue/very high cycle (HCF/VHCF) regime. *Mater. Sci. Eng. A* **2021**, *803*, 140692. [CrossRef]
39. Wang, Q.Y.; Bathias, C.; Kawagoishi, N.; Chen, Q. Effect of inclusion on subsurface crack initiation and gigacycle fatigue strength. *Int. J. Fatigue* **2002**, *24*, 1269–1274. [CrossRef]

9. Zheng, X.T.; Wu, K.W.; Wang, W.; Yu, J.Y.; Xu, J.M.; Ma, L.W. Low cycle fatigue and ratcheting behavior of 35CrMo structural steel at elevated temperature. *Nucl. Eng. Des.* **2017**, *314*, 285–292. [CrossRef]
10. Hua, Z.; Zhang, X.; Zheng, J.; Gu, C.; Cui, T.; Zhao, Y.; Peng, W. Hydrogen-enhanced fatigue life analysis of Cr–Mo steel high-pressure vessels. *Int. J. Hydrogen Energy* **2017**, *42*, 12005–12014. [CrossRef]
11. Wei, W.; Feng, Y.; Han, L.; Zhang, Q.; Zhang, J. Cyclic hardening and dynamic strain aging during low-cycle fatigue of Cr-Mo tempered martensitic steel at elevated temperatures. *Mater. Sci. Eng. A* **2018**, *734*, 20–26. [CrossRef]
12. Ogawa, Y.; Matsunaga, H.; Yamabe, J.; Yoshikawa, M.; Matsuoka, S. Fatigue limit of carbon and Cr Mo steels as a small fatigue crack threshold in high-pressure hydrogen gas. *Int. J. Hydrogen Energy* **2018**, *43*, 20133–20142. [CrossRef]
13. Zhang, J.; Lu, L.; Shiozawa, K.; Zhou, W.; Zhang, W. Effect of nitrocarburizing and post-oxidation on fatigue behavior of 35CrMo alloy steel in very high cycle fatigue regime. *Int. J. Fatigue* **2011**, *33*, 880–886. [CrossRef]
14. Samant, S.; Pandey, V.; Singh, I.; Singh, R. Effect of double austenitization treatment on fatigue crack growth and high cycle fatigue behavior of modified 9Cr–1Mo steel. *Mater. Sci. Eng. A* **2020**, *788*, 139495. [CrossRef]
15. Zhao, Z.L.; Li, C.F. *Metal Technology*; Beijing Institute of Technology Press: Beijing, China, 2019.
16. Shaeri, M.; Saghafian, H.; Shabestari, S. Effect of heat treatment on microstructure and mechanical properties of Cr–Mo steels (FMU-226) used in mills liner. *Mater. Des.* **2012**, *34*, 192–200. [CrossRef]
17. Laxmi, B.; Sharma, S.; Pk, J.; Hegde, A. Quenchant oil viscosity and tempering temperature effect on mechanical properties of 42CrMo4 steel. *J. Mater. Res. Technol.* **2021**, *16*, 581–587. [CrossRef]
18. Gao, C.; Yang, M.; Pang, J.; Li, S.; Zou, M.; Li, X.; Zhang, Z. Abnormal relation between tensile and fatigue strengths for a high-strength low-alloy steel. *Mater. Sci. Eng. A* **2021**, *832*, 142418. [CrossRef]
19. Pang, J.; Li, S.; Wang, Z.; Zhang, Z. General relation between tensile strength and fatigue strength of metallic materials. *Mater. Sci. Eng. A* **2013**, *564*, 331–341. [CrossRef]
20. Pang, J.C.; Li, S.X.; Wang, Z.G.; Zhang, Z.F. Relations between fatigue strength and other mechanical properties of metallic materials. *Fatigue Fract. Eng. Mater. Struct.* **2014**, *37*, 958–976. [CrossRef]
21. Murakami, Y. *Metal Fatigue: Effect of Small Defects and Nonmetallic Inclusions*, 2nd ed.; Elsevier Ltd.: Amsterdam, The Netherlands, 2019.
22. Li, S.X. Effects of inclusions on very high cycle fatigue properties of high strength steels. *Int. Mater. Rev.* **2012**, *57*, 92–114. [CrossRef]
23. Murakami, Y.; Takagi, T.; Wada, K.; Matsunaga, H. Essential structure of S-N curve: Prediction of fatigue life and fatigue limit of defective materials and nature of scatter. *Int. J. Fatigue* **2021**, *146*, 106138. [CrossRef]
24. Liu, Q.; Zhu, G.; Pang, J.; Liu, F.; Li, S.; Guo, C.; Jiang, A.; Zhang, Z. High-cycle fatigue properties prediction and damage mechanisms of RuT400 compacted graphite iron at different temperatures. *Mater. Sci. Eng. A* **2019**, *764*, 138248. [CrossRef]
25. Liu, Y.; Li, Y.; Li, S.; Yang, Z.; Chen, S.; Hui, W.; Weng, Y. Prediction of the S–N curves of high-strength steels in the very high cycle fatigue regime. *Int. J. Fatigue* **2010**, *32*, 1351–1357. [CrossRef]
26. Murakam, Y.; Nomoto, T.; Ueda, T. Factors influencing the mechanism of superlong fatigue failure in steels. *Fatigue Fract. Eng. Mater. Struct.* **1999**, *22*, 581–590. [CrossRef]
27. Duan, Q.Q.; Pang, J.C.; Zhang, P.; Li, S.X.; Zhang, Z.F. Quantitative relations between S-N curves parameters and tensile strength for two steels: AISI 4340 and SCM 435. *Res. Rev. J. Mater. Sci.* **2018**, *6*, 1–16.
28. Shiozawa, K.; Lu, L. Very high-cycle fatigue behaviour of shot-peened high-carbon-chromium bearing steel. *Fatigue Fract. Eng. Mater. Struct.* **2002**, *25*, 813–822. [CrossRef]
29. Petit, J.; Sarrazin-Baudoux, C. An overview on the influence of the atmosphere environment on ultra-high-cycle fatigue and ultra-slow fatigue crack propagation. *Int. J. Fatigue* **2006**, *28*, 1471–1478. [CrossRef]
30. Lee, Y.L.; Pan, J.; Hathaway, R.B.; Barkey, M.E. *Fatigue Testing and Analysis: Theory and Practice*; Butterworth-Heinemann: Waltham, MA, USA, 2005.
31. Gan, Y.; Tian, Z.L.; Dong, H.; Feng, D.; Xin, X.L. *China Materials Engineering Canon Steel Materials Engineering*; Chemical Industry Press: Beijing, China, 2005; Volume 3.
32. Zhao, N.; Zhao, Q.; He, Y.; Liu, R.; Liu, W.; Zheng, W.; Li, L. Strengthening-toughening mechanism of cost-saving marine steel plate with 1000 MPa yield strength. *Mater. Sci. Eng. A* **2021**, *831*, 142280. [CrossRef]
33. Xiao, B.; Xu, L.; Zhao, L.; Jing, H.; Han, Y. Tensile mechanical properties, constitutive equations, and fracture mechanisms of a novel 9% chromium tempered martensitic steel at elevated temperatures. *Mater. Sci. Eng. A* **2017**, *690*, 104–119. [CrossRef]
34. Pang, J.C.; Duan, Q.Q.; Wu, S.D.; Li, S.X.; Zhang, Z.F. Fatigue strengths of Cu-Mg alloy with high tensile strengths. *Scr. Mater.* **2010**, *63*, 1085–1088. [CrossRef]
35. Bayraktar, E.; Garcias, I.; Bathias, C. Failure mechanisms of automotive metallic alloys in very high cycle fatigue range. *Int. J. Fatigue* **2006**, *28*, 1590–1602. [CrossRef]
36. Li, Z.D.; Zhou, S.T.; Yang, C.F.; Yong, Q.L. High/very high cycle fatigue behaviors of medium carbon pearlitic wheel steels and the effects of microstructure and non-metallic inclusions. *Mater. Sci. Eng. A* **2019**, *764*, 138208. [CrossRef]
37. Gao, G.; Liu, R.; Wang, K.; Gui, X.; Misra, R.D.K.; Bai, B. Role of retained austenite with different morphologies on sub-surface fatigue crack initiation in advanced bainitic steels. *Scr. Mater.* **2020**, *184*, 12–18. [CrossRef]

Table 1. *Cont.*

	Dogbone specimens		
No.	Local Stress Amplitude σ_a/MPa	Fatigue Life N/cyc	Loading Frequency f/Hz
1	240	166,822	24
2	300	18,354	4
3	260	127,947	10
4	220	373,935	24
5	350	2704	1
6	350	4156	1
7	260	77,476	10
8	220	2,411,322	24
9	220	691,121	24
10	200	3,000,000 [1]	24
11	200	5,000,000 [1]	24
12	380	2384	1
13	350	4334	1
14	300	31,226	4
15	240	5,000,000 [1]	6~24
16	260	309,419	6~10
17	300	32,059	4
18	260	733,656	10
19	240	10,000,000 [1]	24

[1] Denotes that the specimen does not fail at the associated cycles.

3.3. Crack Initiation Mechanism

Figures 4 and 5 show the SEM images of the fracture surface of several hourglass specimens. It is seen that the fatigue cracks initiate from the specimen surface (Figures 4b and 5b) or the subsurface of the specimen (Figures 4d and 5d). Meanwhile, some specimens exhibit the characteristic of multi-site crack initiation on the fracture surface (Figure 5).

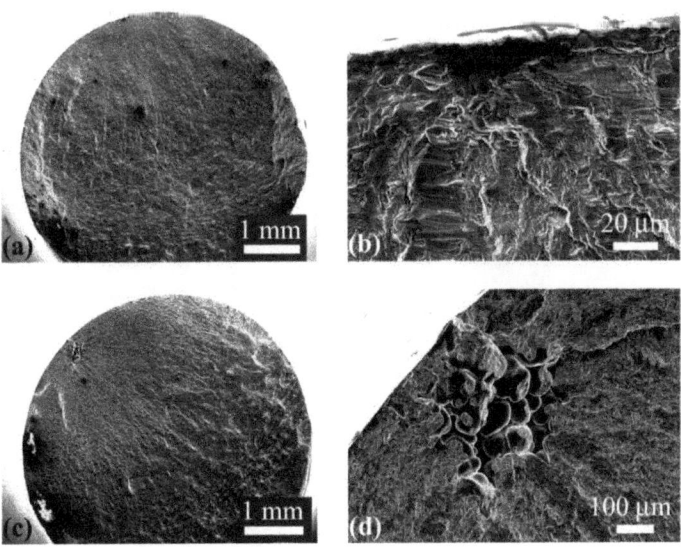

Figure 4. SEM images of the fracture surface for hourglass specimens with single-site crack initiation. (**a**,**b**): local stress amplitude σ_a = 315 MPa, $N = 1.73 \times 10^5$; (**c**,**d**): local stress amplitude σ_a = 252 MPa, $N = 1.37 \times 10^5$.

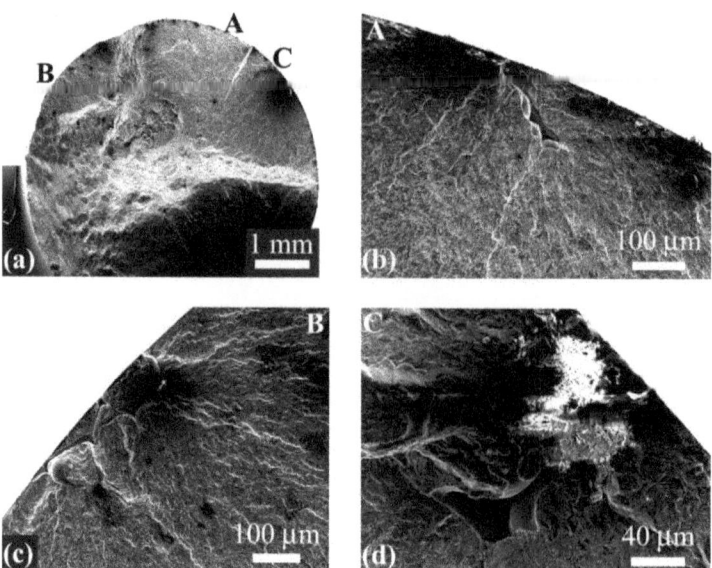

Figure 5. SEM images of the fracture surface for the hourglass specimen with multi-site crack initiation, local stress amplitude σ_a = 315 MPa, $N = 7.96 \times 10^4$. (**a**): Fracture surface with low magnification; (**b**–**d**): close-ups of crack initiation regions A, B, and C in (**a**).

SEM images of the fracture surface of several dogbone specimens are shown in Figures 6 and 7. Similar to hourglass specimens, the fatigue cracks initiate from the specimen surface (Figures 6b and 7b) or the subsurface of the specimen (Figure 6d), and some fracture surfaces present the multi-site crack initiation feature (Figure 7).

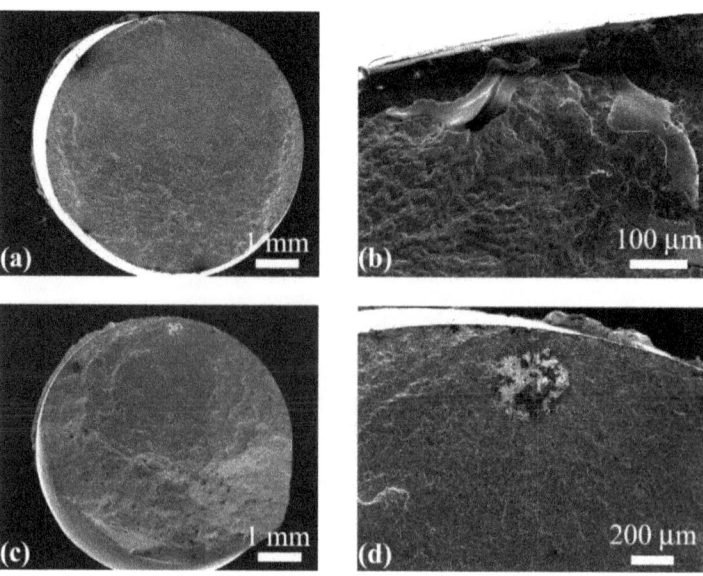

Figure 6. SEM images of the fracture surface for dogbone specimens with single-site crack initiation. (**a**,**b**): σ_a = 220 MPa, $N = 2.41 \times 10^6$; (**c**,**d**): σ_a = 350 MPa, $N = 2.7 \times 10^3$.

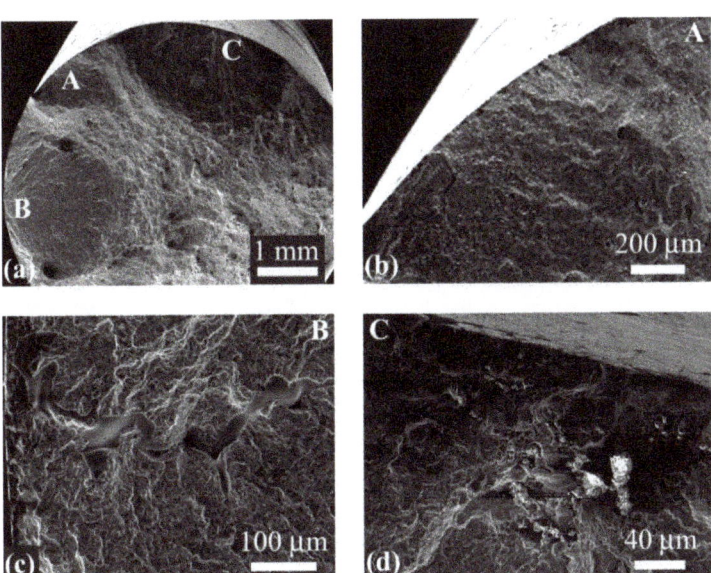

Figure 7. SEM images of the fracture surface for the dogbone specimen with multi-site crack initiation, $\sigma_a = 380$ MPa, $N = 2.38 \times 10^3$. (**a**): Fracture surface with low magnification; (**b**–**d**) Close-ups of crack initiation regions A, B, and C in (**a**).

The SEM observations show that most fatigue cracks initiate from pore defects (Figures 4d, 5c and 6b) or inclusions (Figures 5d and 6d) for both the hourglass specimen and the dogbone specimen, and pore defects are the main crack initiation origins. The specimen size and shape do not change the fatigue failure mechanism of G20Mn5QT steel. The EDS is further used to determine the composition of the inclusion in the crack initiation region. The accelerating voltage is 15 kV. Figure 8 shows the results for the location "+" in the crack initiation region by EDS. It indicates that the main composition of inclusions should be alumina.

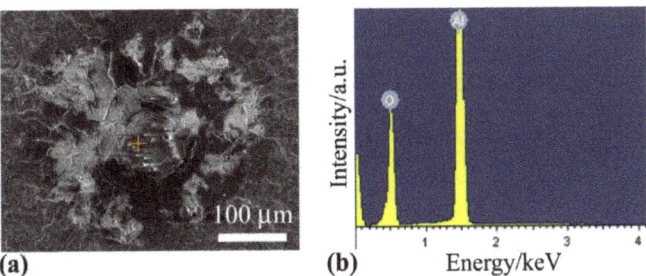

Figure 8. Analysis of composition of the inclusion in crack initiation region in Figure 6d. (**a**) SEM image of the crack initiation region. The symbol "+" denotes the location analyzed by EDS; (**b**) result by EDS.

4. Discussion

4.1. Comparison of S-N Data

Figure 9 shows the comparison of the S-N data between hourglass specimens and dogbone specimens. It is seen from Figure 9a that the difference in the S-N data between the two kinds of specimens is not obvious in terms of nominal stress amplitude, while the fatigue life of the hourglass specimen is generally larger than that of the dogbone specimen for the same local stress amplitude, though the fatigue life data overlap at several low stress amplitudes. As is well-known, the scatter of the fatigue life data tends to be larger at the low stress amplitude (i.e., the long fatigue life). The overlap of the fatigue life data at several low stress amplitudes might be due to the scatter and randomness of the fatigue life. This phenomenon could be explained by the differences among the highly stressed regions of the different types of specimens. The hourglass specimens all fail at or very near the smallest section of the specimen, whereas the positions of the fatigue fracture surface are all located at the parallel segment with the smallest section for the dogbone specimens. The highly stressed region of the hourglass specimen is smaller than that of the dogbone specimen. From the viewpoint of the statistical distribution of microstructures or defects, the dogbone specimen has more possibility for defects or microstructural inhomogeneity that could induce the fatigue failure. This is the reason why the fatigue life of the hourglass specimen is higher than that of the dogbone specimen at the same local stress amplitude. The decrease of the fatigue performance due to the larger highly-stressed region (or control volume) has also been shown for different types of steel in the literature [23,25,26,28–32].

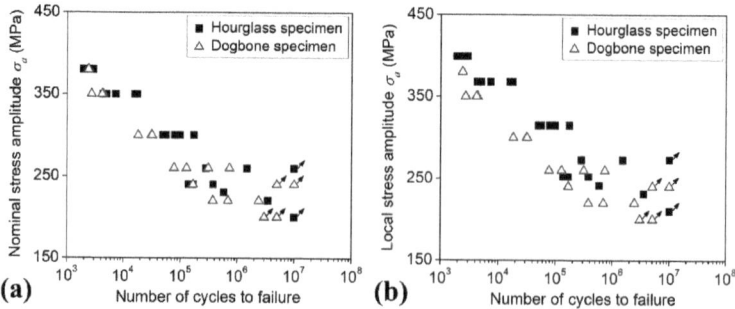

Figure 9. S-N data of specimens with different size and shape, in which the arrows denote the unbroken specimens at the associated cycles. (**a**) Nominal stress amplitude versus fatigue life; (**b**) local stress amplitude versus fatigue life.

4.2. Prediction of Size and Shape Effects

Here, the probabilistic control volume method [25,28] is used to analyze the size and shape effects on the fatigue performance of G20MnQT steel. This method considers that if the fatigue strength of specimens A and B can be regarded as the minimum value of many reference specimens with relatively small control volume under the same manufacturing process and heat treatment, and the fatigue strength of the reference specimen follows a Weibull distribution, the fatigue strength of specimens A and B with the same survival probability satisfies the following relation:

$$\frac{\sigma_A - \gamma}{\sigma_B - \gamma} = \left(\frac{V_A}{V_B}\right)^{-\frac{1}{k}} \quad (1)$$

where σ_A and σ_B denote the fatigue strength of specimens A and B, respectively; V_A and V_B denote the control volume, which is usually chosen as the region with no less than 90% of the maximum principal stress [28–32]; $k > 0$ and $\gamma \geq 0$ are shape and location parameters, respectively.

For the case of fatigue failure induced by the surface crack initiation, the following relation is used:

$$\frac{\sigma_A - \gamma}{\sigma_B - \gamma} = \left(\frac{S_A}{S_B}\right)^{-\frac{1}{k}} \quad (2)$$

where S_A and S_B denote the critical part of the specimen surface (i.e., control surface) with a certain thickness.

In particular, for the two-parameter Weibull distribution, the fatigue strength of specimens A and B at the same survival probabilities satisfies the following relation:

$$\frac{\sigma_A}{\sigma_B} = \left(\frac{V_A}{V_B}\right)^{-\frac{1}{k}} \tag{3}$$

$$\frac{\sigma_A}{\sigma_B} = \left(\frac{S_A}{S_B}\right)^{-\frac{1}{k}} \tag{4}$$

From the consideration that the fatigue cracks initiate from the specimen surface or subsurface for all the hourglass and dogbone specimens, Equation (4) is used to analyze the size and shape effects of the fatigue strength for the present G20Mn5QT steel. The control surface (the region where the principal stress is no less than 90% of the maximum principal stress) is obtained by the finite element analysis. In the calculation, the linear elastic constitutive relation is used. The Young's modulus is $E = 210$ GPa and Poisson's ratio is $\nu = 0.3$. At first, the maximum principal stress is calculated at a load of 100 N under the tensile stress, and then the region of the surface of the specimen where the principal stress is no less than 90% of the maximum principal stress is determined. The control surface of the hourglass and dogbone specimens are listed in Table 2. The parameters of the Weibull distribution of the fatigue strength are estimated by the method in the literature [25,28]. In this method, the bilinear model [25,28,33] is assumed for the S-N curve, i.e.,

$$\log_{10} \sigma = \begin{cases} a \log_{10} N + A, & N < N_0 \\ B, & N \geq N_0 \end{cases} \tag{5}$$

where a, A and B are constants, and N_0 is the number of cycles at the knee point of the curve.

Equation (5) can be written as

$$\log_{10} \sigma = \begin{cases} a(\log_{10} N - \log_{10} N_0) + B, & N < N_0 \\ B, & N \geq N_0 \end{cases} \tag{6}$$

For the specimens with the fatigue strength σ_k and the associated fatigue life N_k ($k = 1, 2, \ldots, n$, and n is the number of specimens), the values of a, B and N_0 can be obtained by the minimum value of the following equation

$$\sum_{N_k < N_0} [\log_{10} \sigma_k - a \log_{10}(N_k/N_0) - B]^2 + \sum_{N_k \geq N_0} (\log_{10} \sigma_k - B)^2 \tag{7}$$

From Equation (6), the fatigue strength σ_k at an arbitrary fatigue life N_k can be transformed into the fatigue strength σ'_k at a given fatigue life N'_k, i.e.,

$$\log_{10} \sigma'_k = \begin{cases} a \log_{10} \frac{N'_k}{N_k} + \log_{10} \sigma_k, & N_k < N_0 \\ a \log_{10} \frac{N'_k}{N_0} + \log_{10} \sigma_k, & N_k \geq N_0 \end{cases} \text{for } N'_k < N_0 \tag{8}$$

or

$$\log_{10} \sigma'_k = \begin{cases} a \log_{10} \frac{N_0}{N_k} + \log_{10} \sigma_k, & N_k < N_0 \\ \log_{10} \sigma_k, & N_k \geq N_0 \end{cases} \text{for } N'_k \geq N_0 \tag{9}$$

Then, the statistical analysis can be performed for the fatigue strength at different fatigue life and the probabilistic stress-life (P-S-N) curve is obtained.

Figure 10 shows the comparison between the predicted P-S-N curves and the experimental data for the hourglass specimen. It is seen that the predicted 50% survival probability curve is in the middle of the experimental data and almost all the experimental data are within the predicted 5% and 95% survival probability curves. This indicates that the predicted results accord well with the experimental data, namely that the method in the literature [25,28] is reasonable for the estimation of the parameters of the Weibull distribution of fatigue strength.

Table 2. Control surface of specimens with different size and shape.

Specimen Type	Hourglass	Dogbone
Control surface/mm^2	73.58	498.14

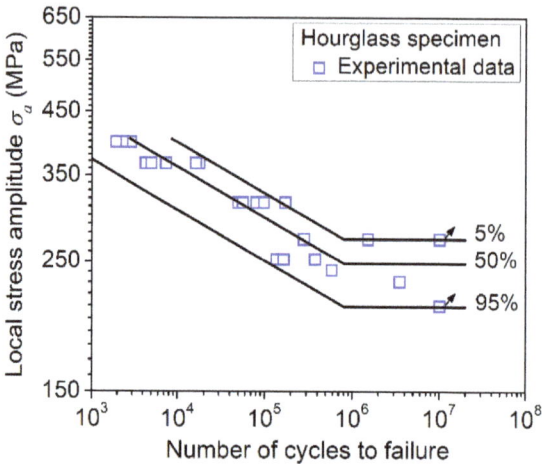

Figure 10. Comparison of predicted P-S-N curves with experimental data for hourglass specimens, in which the arrows denote the unbroken specimens at the associated cycles.

Figure 11 shows the comparison between the predicted P-S-N curves by the experimental data of the hourglass specimen and the experimental data for the dogbone specimen. It is seen that the predicted P-S-N curves are in agreement with the experimental data, indicating that the probabilistic control volume method is applicable for correlating the size and shape effects on the fatigue performance of G20Mn5QT steel.

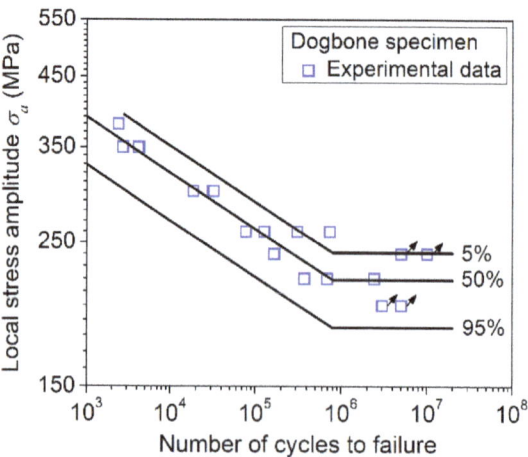

Figure 11. Comparison of predicted P-S-N curves with experimental data of dogbone specimens by using the experimental data of hourglass specimens, in which the arrows denote the unbroken specimens at the associated cycles.

5. Conclusions

In this paper, the size and shape effects on the fatigue behavior are investigated for G20Mn5QT steel of axle box bodies in high-speed trains. The main results are as follows:

The specimen size and shape have influence on the position of the fatigue fracture surface. For the hourglass specimen, it fails at or very near the smallest section of the specimen; whereas for the dogbone specimen, the positions of the fatigue fracture surface are all located at the parallel segment with the smallest section.

The specimen size and shape have no influence on the fatigue failure mechanism of G20Mn5QT steel under an axial loading fatigue test. The fatigue cracks initiate from the surface or the subsurface of the specimen, and some fatigue fracture surfaces exhibit the characteristic of multi-site crack initiation. Most of the fatigue cracks initiate from the pore defects and alumina inclusions in the casting process, and the pore defects are the main crack origins.

The specimen size and shape have an influence on the fatigue performance of G20Mn5QT steel. Due to the larger highly stressed region, the fatigue life of the hourglass specimen is generally higher than that of the dogbone specimen at the same local stress amplitude. The probabilistic control volume method is applicable to correlating the size and shape effects on the fatigue performance of G20Mn5QT steel.

The results are helpful in understanding the fatigue failure mechanism of G20Mn5QT steel and the size and shape effects on the fatigue behavior of metallic materials.

Author Contributions: Conceptualization, Z.Z., Z.L. and C.S.; investigation, Z.Z., Z.L. and C.S.; visualization, Z.L., H.W. and C.S.; writing—original draft preparation, Z.Z., Z.L., H.W. and C.S.; writing—review and editing, Z.Z., Z.L., H.W. and C.S.; supervision, C.S. All authors have read and agreed to the published version of the manuscript.

Funding: This research was funded by the National Key Research and Development Program of China (2017YFB0304600).

Institutional Review Board Statement: Not applicable.

Informed Consent Statement: Not applicable.

Data Availability Statement: Data sharing is not applicable.

Conflicts of Interest: The authors declare no conflict of interest.

References

1. Li, G.; Sun, C.Q. High-temperature failure mechanism and defect sensitivity of TC17 titanium alloy in high cycle fatigue. *J. Mater. Sci. Technol.* **2022**, *122*, 128–140. [CrossRef]
2. Rozumek, D.; Faszynka, S. Surface cracks growth in aluminum alloy AW-2017A-T4 under combined loadings. *Eng. Fract. Mech.* **2020**, *226*, 106896. [CrossRef]
3. Sun, C.Q.; Li, Y.Q.; Xu, K.L.; Xu, B.T. Effects of intermittent loading time and stress ratio on dwell fatigue behavior of titanium alloy Ti-6Al-4V ELI used in deep-sea submersibles. *J. Mater. Sci. Technol.* **2021**, *77*, 223–236. [CrossRef]
4. Rozumek, D. Influence of the slot inclination angle in FeP04 steel on fatigue crack growth under tension. *Mater. Des.* **2009**, *30*, 1859–1865. [CrossRef]
5. Lukács, J.; Meilinger, Á.; Pósalaky, D. High cycle fatigue and fatigue crack propagation design curves for 5754-H22 and 6082-T6 aluminium alloys and their friction stir welded joints. *Weld. World* **2018**, *62*, 737–749. [CrossRef]
6. Mädler, K.; Geburtig, T.; Ullrich, D. An experimental approach to determining the residual lifetimes of wheelset axles on a full-scale wheel-rail roller test rig. *Int. J. Fatigue* **2016**, *86*, 58–63. [CrossRef]
7. Wang, Y.; Yuan, L.; Zhang, S.; Sun, C.; Wang, W.; Yang, G.; Wei, Y. The influence of combined gradient structure with residual stress on crack-growth behavior in medium carbon steel. *Eng. Fract. Mech.* **2019**, *209*, 369–381. [CrossRef]
8. Wu, S.C.; Liu, Y.X.; Li, C.H.; Kang, G.Z.; Liang, S.L. On the fatigue performance and residual life of intercity railway axles with inside axle boxes. *Eng. Fract. Mech.* **2018**, *197*, 176–191. [CrossRef]
9. Akama, M. Bayesian analysis for the results of fatigue test using full-scale models to obtain the accurate failure probabilities of the Shinkansen vehicle axle. *Reliab. Eng. Syst. Safe* **2002**, *75*, 321–332. [CrossRef]
10. Xu, Y.H.; Zhao, Y.X. Simulation for short fatigue cracks initiation of LZ50 axle steel for railway vehicles. *Chin. J. Appl. Mech.* **2009**, *26*, 589–593.

11. Cervello, S. Fatigue properties of railway axles: New results of full-scale specimens from Euraxles project. *Int. J. Fatigue* **2016**, *86*, 2–12. [CrossRef]
12. Lu, L.T.; Zhang, J.W.; Zhang, Y.B.; Zhi, B.Y.; Zhang, W.H. Rotary bending fatigue property of LZ50 axle steel in gigacycle regime. *J. China Railw. Soc.* **2009**, *31*, 37–41.
13. Chen, Y.P.; Li, Y.B.; Zhang, X.L.; Sun, C.Q.; Hong, Y.S. Study on high-cycle and very-high-cycle fatigue properties of EA4T axle steel. *Rail Transp. Equip. Technol.* **2017**, *1*, 21–23.
14. Beretta, S.; Carboni, M.; Fiore, G.; Lo Conte, A. Corrosion–fatigue of A1N railway axle steel exposed to rainwater. *Int. J. Fatigue* **2010**, *32*, 952–961. [CrossRef]
15. Wang, W.J.; Liu, Z.M.; Li, Q.; Miao, L.X. Fatigue strength analysis of CRH2 motor bogie frame. *J. Beijing Jiaotong Univ.* **2009**, *33*, 5–9.
16. Zhang, S.J.; Xie, J.J.; Jiang, Q.Q.; Zhang, X.L.; Sun, C.Q.; Hong, Y.S. Fatigue crack growth behavior in gradient microstructure of hardened surface layer for an axle steel. *Mater. Sci. Eng. A* **2017**, *700*, 66–74. [CrossRef]
17. Guagliano, M.; Vergani, L. Experimental and numerical analysis of sub-surface cracks in railway wheels. *Eng. Fract. Mech.* **2005**, *72*, 255–269. [CrossRef]
18. Gao, J.W.; Pan, X.N.; Han, J.; Zhu, S.P.; Liao, D.; Li, Y.B.; Dai, G.Z. Influence of artificial defects on fatigue strength of induction hardened S38C axles. *Int. J. Fatigue* **2020**, *139*, 105746. [CrossRef]
19. Luke, M.; Varfolomeev, I.; Lütkepohl, K.; Esderts, A. Fatigue crack growth in railway axles: Assessment concept and validation tests. *Eng. Fract. Mech.* **2011**, *78*, 714–730. [CrossRef]
20. Tomaszewski, T.; Sempruch, J. Analysis of size effect in high-cycle fatigue for EN AW-6063. *Solid State Phenom.* **2014**, *224*, 75–80. [CrossRef]
21. Lee, Y.L.; Pan, J.; Hathaway, R.; Barkey, M. *Fatigue Testing and Analysis: Theory and Practice*; Elsevier Butterworth-Heinemann: Oxford, UK, 2005.
22. Li, Y.Q.; Song, Q.Y.; Feng, S.C.; Sun, C.Q. Effects of loading frequency and specimen geometry on high cycle and very high cycle fatigue life of a high strength titanium alloy. *Materials* **2018**, *11*, 1628. [CrossRef] [PubMed]
23. Furuya, Y. Specimen size effects on gigacycle fatigue properties of high-strength steel under ultrasonic fatigue testing. *Scr. Mater.* **2008**, *58*, 1014–1017. [CrossRef]
24. Shirani, M.; Härkegård, G. Fatigue life distribution and size effect in ductile cast iron for wind turbine components. *Eng. Fail. Anal.* **2011**, *18*, 12–24. [CrossRef]
25. Li, Y.B.; Song, Q.Y.; Yang, K.; Chen, Y.P.; Sun, C.Q.; Hong, Y.S. Probabilistic control volume method for the size effect of specimen fatigue performance. *Chin. J. Theor. App. Mech.* **2019**, *51*, 1363–1371. [CrossRef]
26. Shen, X.L.; Lu, L.T.; Jiang, H.F.; Zhang, J.W.; Yi, H.F. Effect of inclusion size on the fatigue strengthen of small specimens and railway alloy axles. *J. Mech. Eng.* **2010**, *46*, 48–52. [CrossRef]
27. Varfolomeev, I.; Luke, M.; Burdack, M. Effect of specimen geometry on fatigue crack growth rates for the railway axle material EA4T. *Eng. Fract. Mech.* **2011**, *78*, 742–753. [CrossRef]
28. Sun, C.Q.; Song, Q.Y. A method for predicting the effects of specimen geometry and loading condition on fatigue strength. *Metals* **2018**, *8*, 811. [CrossRef]
29. Murakami, Y. *Metal Fatigue: Effects of Small Defects and Nonmetallic Inclusions*; Elsevier Science Ltd.: Oxford, UK, 2002; pp. 333–336.
30. Sun, C.Q.; Zhang, X.; Liu, X.; Hong, Y.S. Effects of specimen size on fatigue life of metallic materials in high-cycle and very-high-cycle fatigue regimes. *Fatigue Fract. Eng. Mater. Struct.* **2016**, *39*, 770–779. [CrossRef]
31. Sun, C.Q.; Song, Q.Y. A method for evaluating the effects of specimen geometry and loading condition on fatigue life of metallic materials. *Mater. Res. Express* **2019**, *6*, 046536. [CrossRef]
32. Li, C.M.; Hu, Z.; Sun, C.Q.; Song, Q.Y.; Zhang, W.H. Probabilistic control volume method for evaluating the effects of notch size and loading type on fatigue life. *Acta Mech. Solida Sin.* **2020**, *33*, 141–149. [CrossRef]
33. Hanaki, S.; Yamashita, M.; Uchida, H.; Zako, M. On stochastic evaluation of S–N data based on fatigue strength distribution. *Int. J. Fatigue* **2010**, *32*, 605–609. [CrossRef]

Communication

Revealing the Formation of Recast Layer around the Film Cooling Hole in Superalloys Fabricated Using Electrical Discharge Machining

Zenan Yang [1], Lu Liu [2], Jianbin Wang [2], Junjie Xu [1], Wanrong Zhao [1], Liyuan Zhou [1], Feng He [2] and Zhijun Wang [2,*]

1. Science and Technology on Advanced High-Temperature Structural Materials Laboratory, Beijing Institute of Aeronautical Materials, Beijing 100095, China
2. State Key Laboratory of Solidification Processing, Northwestern Polytechnical University, Xi'an 710072, China
* Correspondence: zhjwang@nwpu.edu.cn; Tel.: +86-134-8467-1484

Abstract: A film cooling hole is an efficient and reliable cooling method, which is widely used in aeroengine turbine blades to effectively improve the thrust–weight ratio of the engine. Electrical discharge machining is the most common manufacturing process for film cooling holes. Due to the rapid quenching after high-temperature melting, a certain thickness of the recast layer will be formed in the vicinity of the hole wall. The microstructure of the recast layer is considered to be an important factor affecting the performance of single-crystal blades. Generally, the recast layer has been thought of as one of the main reasons for the failure of turbine blades. Accordingly, the formation of the recast layer is an important and interesting issue to be revealed. In this work, the recast layer formed using electrical discharge machining on a single-crystal superalloy is studied with TEM. It is found that the recast layer is in the state of supersaturated solution with a single-crystal structure epitaxially grown from the matrix, and many dislocations were observed therein.

Keywords: film cooling hole; electrical discharge machining; recast layer; microstructure

1. Introduction

Along with the increasing demand for turbine inlet temperature of aeroengines, the thermal barrier coatings and film cooling holes on single-crystal, nickel-based superalloys blades have become an effective way to improve cooling efficiency [1–3]. Film cooling technology is an important innovation that was first applied to the anti-icing of aircraft wings [4]. The discrete holes on the blade's surface with cooling gas passing through can isolate the blade's surface from high temperatures, playing a dual role in heat insulation and cooling. The film cooling holes distributed in turbine blades of aeroengines have several typical characteristics [5]. The aperture of the film cooling hole is very small, approximately in the region of 0.25~1.25 mm. Film cooling holes at different positions may have different crystalline directions. There are also a large number of film holes in a single turbine blade, which may be more than thousands. The distribution characteristics of film cooling holes in the blade make it difficult for the traditional forming processes to meet the requirements. The formation of the recast layer (RL) in the electrical discharge machining (EDM) process has been paid attention to for a long time since it has been thought of as one of the main reasons for the fracture of film cooling holes.

The processing of film cooling holes mainly relies on special processing methods such as EDM, laser drilling (LD) and electro-chemical machining (ECM). Up to now, EDM was the preferred method for manufacturing film cooling holes [6]. The EDM drilling process is mainly composed of a rotating hollow tubular electrode, a high-voltage working fluid and a machined workpiece [7]. The processing principle is to use the pulse discharge between the electrodes to etch the workpiece material while introducing a high-voltage working fluid into the tubular electrode to flush away the machining debris and ensure the normal discharge of the next pulse [8]. Due to the thermal etching process in EDM, most of the melted and vaporized metals during the thermal etching stage are thrown into the coolant and become small particles, while the remaining part is rapidly cooled and resolidified on the wall and blades surface to form the RL. There is also an adjacent heat-affected zone (HAZ) below the RL [9]. The formation of the RL in the EDM process has been paid attention to for a long time [10,11] since it has been thought of as one of the main reasons for the fracture of film cooling holes.

Previous investigations focused on the factors affecting the recast layer, including processing parameters and the different types of materials. Using the thermal–thermal coupling model, Tang et al. [12,13] reported that most molten metal would remain in the spark pits during a single discharge process to form the recast layer. The material in the process greatly affects the structure of the recast layer. Cusanelli et al. [14] found that due to the presence of carbon, the RL mainly consisted of residual austenite and columnar martensite, and the hardness of the RL was twice that of the ferritic matrix for EDM on W300 steel. Murray et al. [15] found that the RL in monocrystalline silicon was composed of two crystal grains with an amorphous phase. Liu et al. [16] investigated the composition and microstructure of the RL on a nickel–titanium memory alloy and identified that the RL produced using EDM alloy was crystalline.

For the recast layer around the cooling hole of the single-crystal superalloy, the IN738 superalloy, as one of the representative nickel-based precipitation hardening alloys, tends to form a single-crystal structure, and the composition distribution is uniform [17]. Dong et al. [18] investigated two single-crystal superalloys, IC21 and N5, for EDM and found that the composition of the RL can be considered as an imperfect single-crystal structure. On this basis, Shang et al. [19] further confirmed that the RL does not contain a precipitated strengthening phase, and the microhardness is smaller than that of the matrix. The rapid melting and solidification process will lead to defect formation in single-crystal superalloys. In the present work, the superalloys used in turbine blades have gradually developed into the fourth generation with increasing volume fraction of precipitation. A deeper observation at the nanoscale is still limited, and the formation of the RL in alloys with high-volume fraction precipitation also needs to be revealed. This work introduces high-resolution transmission electron microscopy to clarify the nanostructure of the recast layer.

2. Experimental Methods

The single-crystal superalloy used in this study is a fourth-generation superalloy with an approximately 70% volume fraction of precipitation. The chemical composition of the alloy is shown in Table 1.

Table 1. Nominal chemical composition of the fourth-generation, nickel-based, single-crystal superalloy (mass fraction/%) [20].

Element	Cr	Co	Mo	W	Ta	Re	Ru	Nb	Al	Hf	C	Ni
mass fraction/%	2.0~4.0	7.0~10.0	0.8~1.6	6.0~8.0	7.0~9.0	3.0~5.0	2.0~4.0	0.2~1.0	5.0~6.0	0.1~0.3	0.008	Bal.

The film cooling holes were machined on the single-crystal superalloy using EDM. The machining diameter of the film cooling hole is $\Phi = 0.5$ mm, the hole depth is 2 mm, and the hole axis is perpendicular to the surface of the sample. Among the adjustable parameters, the regulation range of pulse width in the electrical parameters is 3~11 μs, the pulse interval is between 8 and 16 μs and the peak current is regulated between 2 and 11 A. The electrode speed and flushing pressure in the non-electric parameters are regulated between 50 and 250 rpm and 0.2 and 0.6 MPa, respectively. The influence of process parameters on the processing quality and efficiency of film cooling holes is different. When the pulse width is 4 μs, the pulse interval is 16 μs, the peak current is 4 A, the electrode rotation speed is 150 rpm and the flushing pressure is 0.4 MPa, the film cooling hole with better processing quality can be obtained, and, especially, the thickness of the recast layer is thinner and uniform.

Due to the small range of the recast layer, the transmission electron microscopy (TEM) samples were prepared using a focused ion beam (FIB) in SEM (FEI Helios G4 CX). Firstly, the film cooling hole with a thick recast layer was selected with SEM, and then the sample was cut around the hole where the recast layer thickness was uniform. The nanostructure is observed with high-resolution TEM (HR-TEM) of JEM-ARM300F. Figure 1 shows the procedures of the initial lift-out and final thinning. Selecting the specified region is shown in Figure 1a,b. A Ga ion beam is used to mill away trenches adjacent to the region, as shown in Figure 1c. The sample then was handled and welded to a prefabricated Cu-grid and thinned from each side to the desired thickness, as shown in Figure 1d. The range of the recast layer and matrix in the sample is shown in Figure 1d. The clear boundary between the RL and the matrix can be seen in Figure 1.

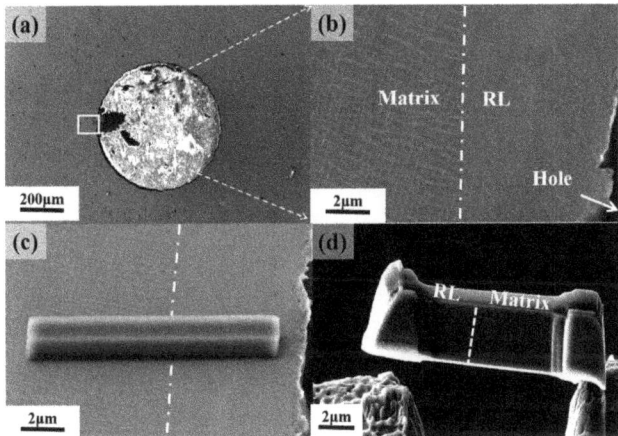

Figure 1. TEM sample preparation: (**a**) The selected specified region with both the matrix and recast layer. (**b**) Microstructure of the selected region. (**c**) The exact location of the TEM sample. (**d**) The prepared sample using FIB. The recast layer (left) and matrix (right) are shown in the sample.

The hardness difference between the recast layer and matrix was investigated using a Hysitron Ti950 nanoindentation with a Berkovich tip, which was calibrated on a standardized fused quartz specimen. The loading and saturation times were set to 5 s and 2 s, respectively. In the side wall of the film cooling hole along the matrix direction, at every 8 μm, a point was pressured, and in the vertical direction of the hole side wall, at every 5 μm, a point was pressured, for a total of 6 points.

3. Results of the Characterizations of the Recast Layer

Figure 2 shows a scanning transmission electron microscopy high angle annular dark field (STEM-HAADF) image of the sample. There are three different regions, which are marked as RL, RL–matrix, and matrix in Figure 2a, which were selected for further TEM analyses. It is obvious that the recast layer and the matrix show a significantly different microstructure. In the matrix, the typical γ'-cube precipitation is regularly distributed in the γ phase, while in the recast layer, the uniform image indicates a single-phase structure without any precipitation. As illustrated in Figure 2b,c, both the dark field (DF) and bright field (BF) images show a large number of dislocations distributing in the recast layer, while the matrix consists of the γ channel and cubical γ' phase (Figure 2f,g). A clear boundary can be detected between the RL and the matrix, as shown in Figure 2d,e.

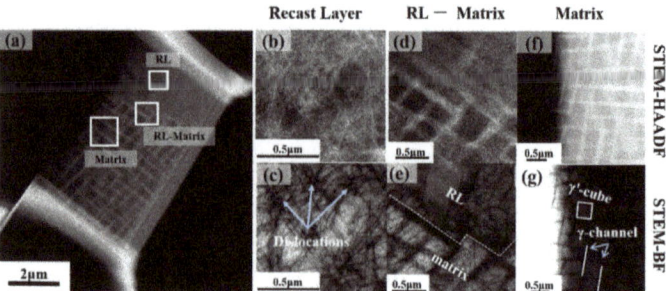

Figure 2. The detailed TEM analysis for the prepared sample: (**a**) The STEM-HAADF image of the overall microstructure. STEM-HAADF and STEM-BF images of (**b**,**c**) the recast layer, (**d**,**e**) the recast layer (RL) and matrix transition area and (**f**,**g**) the matrix of the sample. High-density dislocations are distributed over the area of the recast layer, while the matrix shows a typical γ'-γ microstructure.

To further identify the phase compositions of the two regions, the selected area electron diffraction (SAED) patterns of the recast layer and the matrix are presented in Figure 3. The diffraction patterns in the recast layer and the matrix were obtained under the same tilting condition. The BF image and corresponding SAED pattern in Figure 3a,b indicate that the RL consists of only a single FCC phase. Meanwhile, the SAED pattern in the matrix in Figure 3c,d shows the superlattice of the $L1_2$ phase, corresponding to the γ' precipitation. Combined with Figure 2, it reveals that after high-temperature melting and rapid quenching, the epitaxial growth of molten metal occurs from the single-crystal matrix, and the same crystal orientation is maintained as the same as the matrix. The excessive cooling rate prevents the precipitation of the γ' phase during the rapid solidification process. The recast layer continuously formed along the matrix with epitaxial growth is related to the continuous molten pool and temperature gradient generated during EDM. Each molten pool formed in the drilling process can be considered distributed perpendicular to the inner surface of the hole. This means that the molten pool is distributed in the same plane around the hole along the axial direction on the cross-section of the recast layer. The directional temperature gradient of the molten pool ensures the single-crystal properties of the matrix and the recast layer after processing.

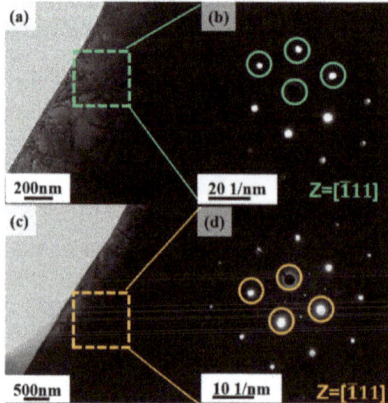

Figure 3. Selected area electron diffraction patterns of (**a**,**b**) the recast layer and (**c**,**d**) the matrix. The recast layer has a single-crystal structure with the same orientation in matrix.

The composition of the recast layer and matrix were also identified with EDS in TEM. As shown in Figure 4, Cr, Co and Re are concentrated in the γ phase of the selected matrix region, while this phenomenon does not exist in the recast layer region. The other elements do not vary significantly between the two regions. It is also shown that there is no compositional segregation in the recast layer, which further proves that no new phase is generated during the formation of the recast layer, which is consistent with the STEM and SAED analysis. The homogeneous redistribution of the elements in the recast layer is related to the rapid solidification experienced by the alloy in the process of electrical discharge. The alloy melts after reaching the melting point, and the elements are redistributed uniformly in the liquid. Then, the fast cooling leads to the absence of segregation, and there is no solute diffusion in the liquid. Accordingly, the recast layer only forms a single γ phase.

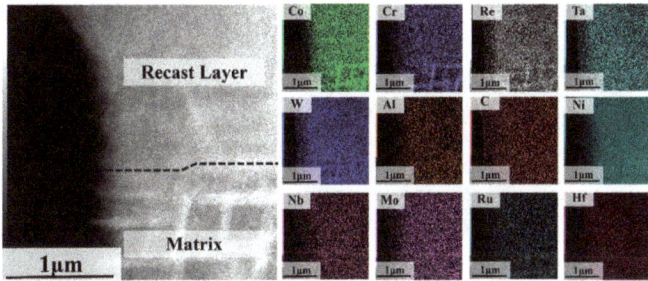

Figure 4. Energy dispersive spectroscopy area scanning profiles of the recast layer and matrix transition region. The composition of the recast layer area is uniform, and there is no segregation for all the elements.

Figure 5 shows the high-resolution TEM (HR-TEM) images and geometric phase analysis of the recast layer and the matrix. The lattice spacing of ($\bar{1}11$) plane is estimated as 0.2099 nm in the recast layer, which is slightly larger than the lattice spacing of the FCC matrix (Figure 5(a1,a2)). The larger lattice constant of the recast layer is due to the supersaturation of elements with larger atomic sizes. The alloying elements with large atomic radii are dissolved in the recast layer, resulting in a larger lattice constant than that of the matrix. This is consistent with the results shown in Figure 4.

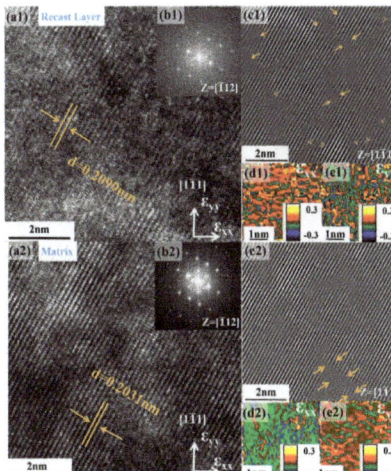

Figure 5. High-resolution transmission electron microscopy analyses of the recast layer (**a1–e1**) and the matrix (**a2–e2**). (**a1,a2**) HR-TEM atomic interface topographies. (**b1,b2**) Fast Fourier transform images. (**c1,c2**) Inverse fast Fourier transform images. (**d1,d2**) Geometric phase analysis (GPA) in the x direction. (**e1,e2**) Geometric phase analysis in the y direction. The atomic distortion and dislocation distribution in the recast layer region is more obvious than in the matrix.

Using geometric phase analysis (GPA), it is found that there is a wide range of strain concentration areas in the given x and y directions of the recast layer (Figure 5(d1,e1)), which is due to the high density of dislocations in the recast layer (Figure 5(c1)). Moreover, the microscopic strain of the matrix itself is less obvious (Figure 5(d2,e2)). The differences in lattice constants and microscopic strains between the recast layer and the matrix are closely related to the large number of dislocation defects caused by residual stress during machining, i.e., during the formation of the recast layer, it will be affected by multiple pulse discharges. With the increase in discharge times, the internal residual stress will gradually increase. In the process of forming the recast layer, there will be more dislocation density, resulting in more obvious distortion in the crystal lattice. This is one of the important reasons why there are more strain concentration areas in the recast layer in nanoscale GPA analyses.

According to the analyses shown in Figure 2b,e, under the STEM of the FIB sample, there is a clear boundary between the recast layer and the matrix transition interface, which indicates that the transition area is very small. In order to determine the range of the heat-affected zone more accurately, the diffraction pattern difference and lattice constant change in the recast layer and the matrix on both sides of the interface are explored using a high-resolution transmission image. Figure 6a shows the high-resolution transmission image (view field of 35 nm × 35 nm) near the transition region in the FIB sample. Figure 6b,d shows the corresponding recast layer and matrix regions. Figure 6c,e shows the corresponding diffraction patterns in different regions. It reveals that the distance between the [$\bar{1}12$] atomic plane in the recast layer is larger than that in the matrix. The position of the transition interface can be determined in this way. At the same time, the microstrain inside the heat-affected zone is analyzed. The geometric phase analysis of the microscopic strain distribution in the transition region is further shown in Figure 6f,g. There are differences in the microscopic strain distribution in the transition interface in the x and y directions. The strain in the x direction near the matrix is mainly compressive strain, and the strain in the x direction near the recast layer is mainly tensile strain. This is mainly caused by the difference in the lattice constant of the recast layer and the matrix. The strain on both sides of the y-axis direction is tensile strain. The crystal structure in both the recast layer and the matrix is FCC, but the lattice constant of the recast layer is larger than that of the matrix due to the single-phase structure and uniform distribution of elements. The strain in the x-direction difference in the transition region indicates the difference in the atomic arrangement between the recast layer and the matrix and explains the existence of a certain degree of misorientation between the recast layer and the matrix.

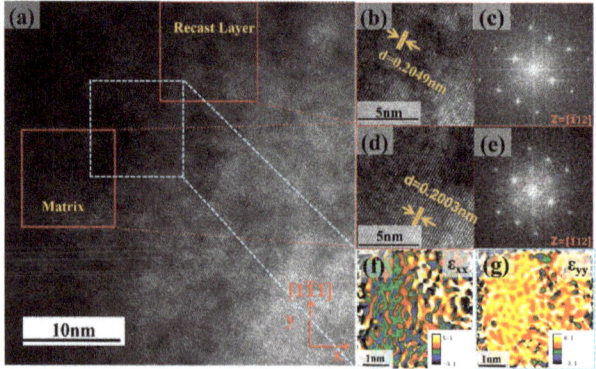

Figure 6. HR-TEM analysis of transition regions: (**a**) HR-TEM image; (**b**,**c**) HR-TEM and FFT of the recast layer; (**d**,**e**) HR-TEM and FFT of the matrix; and (**f**,**g**) geometric phase analysis of transition regions along the x and y directions.

4. Discussion

As the main result of the rapid melting and solidification process, the recast layer has many unique characteristics in terms of the forming method and structure compared with other materials. When the high-power energy of the single-crystal superalloy is given instantaneously, a directional temperature gradient will appear in the molten pool. The recast layer will achieve epitaxial growth along the direction of the single-crystal matrix during the high-temperature melting and rapid solidification process, maintain a continuous crystal orientation and finally grow into a single-crystal structure connected to the matrix material. For the discharge generated with multiple pulse cycles, the overlap and intersection of the molten pool further led to the emergence of the low-angle boundary. The nickel-based, single-crystal superalloy is mainly composed of the γ phase and γ' phase.

The formation of the recast layer is a result of rapid melt–solidification, a typical process with many special features. Figure 7 schematically shows the formation processes of the recast layer on the single-crystal superalloy. The pulse duration of EDM is very short. However, the highest temperature in the discharge process can increase rapidly over the melting point of the alloy. The γ phase and ordered $L1_2$ phase melt to form a liquid film. After the end of discharge, the cooling rate is more than 10^6 K/s, and the molten metal solidifies rapidly. The accelerated cooling rate prevents precipitation. In this process, the higher directional temperature gradient field makes the recast layer epitaxially grow along the matrix. At the same time, severe thermal stress generates, resulting in a much higher dislocation density in the recast layer region than that in the matrix region.

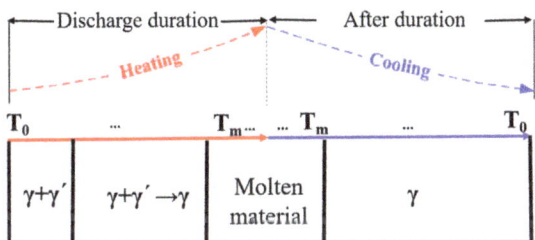

Figure 7. Schematic diagram of recast layer evolution of the fourth-generation single-crystal superalloy after EDM.

The recast layer around the cooling hole no longer contains the precipitated strengthening phase γ', so the hardness may be lower than the matrix with precipitation.

Figure 8 shows the nanoindentation hardness of the film cooling hole wall at different locations along the hole wall–matrix direction (x direction) and the hole inlet–hole outlet direction (y direction). In Figure 8a, the position far away from the hole wall is larger than 7 GPa. The three testing points in Figure 8b are all located on the recast layer, and the hardness for all is smaller than 5 GPa. Compared with the hardness, point 1 in Figure 8a at 3 μm near the hole wall may be adjacent to the recast layer. The nanohardness of point 2 and point 3 in Figure 8a is not much different, and both are greater than point 1. Moreover, point 2 and point 3 are 11 μm and 19 μm away from the hole wall, respectively, and could be confirmed on the matrix due to the small range of the heat-affected zone [19]. Based on Figure 8a,b, it can be considered that the recast layer of the hole wall is much softer than the nearby alloy matrix. In addition, the nanohardness is lower than 5 GPa, while the hardness of the matrix in Figure 8a is greater than 7 GPa. The hardness of the recast layer is lower than that of the matrix, which agrees with the microstructure analysis above. The molten metal is remelted and solidified to adhere to the hole wall, and the recast layer is determined to be a single-phase γ structure using various analysis methods and does not contain precipitated strengthening γ' phase, so the overall hardness value is low.

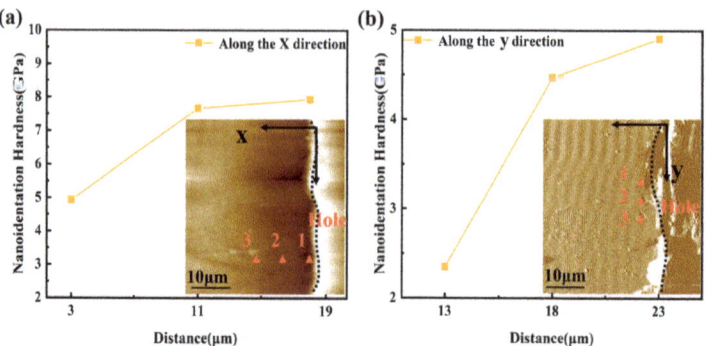

Figure 8. Nanoindentation hardness of the recast layer and matrix at different positions. (**a**) Schematic diagram of nanoindentation hardness and indentation of the recast layer and matrix in the x direction. (**b**) Schematic diagram of nanoindentation hardness and indentation of the recast layer in the y direction. The hardness of the recast layer is much lower than that of the matrix, caused by the lack of precipitation.

Considering the varied nanohardness of the recast layer at different positions, the different mechanical behavior between the recast layer and the matrix may be a weakness at the crack initiation point. In the future, it is necessary to further confirm the hardness evolution of the recast layer during service. Moreover, the different oxidation behavior may be exhibited for different thicknesses of the recast layer.

5. Conclusions

In this work, the microstructure and crystal structure of the recast layer of a single-crystal superalloy after EDM were studied using FIB and HR-TEM. The recast layer is formed by a localized molten pool induced with drilling, which undergoes directional solidification with a specific temperature gradient, and epitaxial growth occurs along the single-crystal alloy matrix. The recast layer has a single-phase structure composed of a supersaturated γ phase with the same orientation in the matrix. In the recast layer, there are plenty of defects, and its lattice constant is larger than that in the matrix due to the supersaturated solutions. It is determined that the recast layer has the characteristics of a rapid solidification nonequilibrium structure. There is a small mismatch between the recast layer and the matrix. The nano-indentation test confirms that the hardness of the recast layer near the hole wall is lower than that of the nickel-based, single-crystal superalloy substrate, which might affect the creep performance of the blade. The recast layer distributed on the film cooling hole wall easily falls off the surface of the alloy due to the difference in the formation mode and the structure of the matrix.

Author Contributions: Conceptualization, Z.Y.; data Curation, L.L. and J.W.; formal analysis, Z.Y. and L.L.; resources, Z.Y.; writing—original draft preparation, Z.Y. and L.L.; supervision, J.W. and J.X.; validation, W.Z. and L.Z.; writing—review and editing, F.H. and Z.W.; project administration, Z.W. All authors have read and agreed to the published version of the manuscript.

Funding: The authors are grateful for support from the Research Fund of the State Key Laboratory of Solidification Processing (NPU), China (Grant No. 2022-BJ-03).

Data Availability Statement: The data presented in this study are available on request from the corresponding author.

Conflicts of Interest: The authors declare that they have no conflict of interest.

References

1. Miller, R.A. Thermal barrier coatings for aircraft engines: History and directions. *J. Therm. Spray Technol.* **1997**, *6*, 35–42. [CrossRef]
2. He, K. Investigations of film cooling and heat transfer on a turbine blade squealer tip. *Appl. Therm. Eng.* **2017**, *110*, 630–647.
3. Liu, Y.; Ru, Y.; Zhang, H.; Pei, Y.; Li, S.; Gong, S. Coating-assisted deterioration mechanism of creep resistance at a nickel-based single-crystal superalloy. *Surf. Coat. Technol.* **2021**, *406*, 126668. [CrossRef]
4. Wieghardt, K. *Hot-Air Discharge for De-Icing*; Wright Field, Air Materiel Command: Dayton, OH, USA, 1946; pp. 1–44.
5. Bunker, R.S. A Review of Shaped Hole Turbine Film-Cooling Technology. *J. Heat Transf. Trans. Asme* **2005**, *127*, 441–453. [CrossRef]
6. Ay, M.; Çaydaş, U.; Hasçalık, A. Optimization of micro-EDM drilling of inconel 718 superalloy. *Int. J. Adv. Manuf. Technol.* **2013**, *66*, 1015–1023. [CrossRef]
7. Tao, X.; Liu, Z.; Qiu, M.; Tian, Z.; Shen, L. Research on an EDM-based unitized drilling process of TC4 alloy. *Int. J. Adv. Manuf. Technol.* **2018**, *97*, 867–875. [CrossRef]
8. Li, C.; Xu, X.; Li, Y.; Tong, H.; Ding, S.; Kong, Q.; Zhao, L.; Ding, J. Effects of dielectric fluids on surface integrity for the recast layer in high speed EDM drilling of nickel alloy. *J. Alloys Compd.* **2019**, *783*, 95–102. [CrossRef]
9. Pei, H.Q.; Wang, J.J.; Li, Z.; Li, Z.W.; Yao, X.Y.; Wen, Z.X.; Yue, Z.F. Oxidation behavior of recast layer of air-film hole machined by EDM technology of Ni-based single crystal blade and its effect on creep strength. *Surf. Coat. Technol.* **2021**, *419*, 127285. [CrossRef]
10. Kuppan, P.; Rajadurai, A.; Narayanan, S. Influence of EDM process parameters in deep hole drilling of Inconel 718. *Int. J. Adv. Manuf. Technol.* **2008**, *38*, 74–84. [CrossRef]
11. Wang, F.; Liu, Y.; Shen, Y.; Ji, R.; Tang, Z.; Zhang, Y. Machining Performance of Inconel 718 Using High Current Density Electrical Discharge Milling. *Mater. Manuf. Process.* **2013**, *28*, 1147–1152. [CrossRef]
12. Tang, J.; Yang, X. A novel thermo-hydraulic coupling model to investigate the crater formation in electrical discharge machining. *J. Phys. D Appl. Phys.* **2017**, *50*, 365301. [CrossRef]
13. Tang, J.; Yang, X. Simulation investigation of thermal phase transformation and residual stress in single pulse EDM of Ti-6Al-4V. *J. Phys. D Appl. Phys. A Europhys. J.* **2018**, *51*, 135308. [CrossRef]
14. Cusanelli, G.; Hessler-Wyser, A.; Bobard, F.; Demellayer, R.; Perez, R.; Flükiger, R. Microstructure at submicron scale of the white layer produced by EDM technique. *J. Mater. Process. Technol.* **2004**, *149*, 289–295. [CrossRef]
15. Murray, J.W.; Fay, M.W.; Kunieda, M.; Clare, A.T. TEM study on the electrical discharge machined surface of single-crystal silicon. *J. Mater. Process. Technol.* **2013**, *213*, 801–809. [CrossRef]
16. Liu, J.F.; Guo, Y.B.; Butler, T.M.; Weaver, M.L. Crystallography, compositions, and properties of white layer by wire electrical discharge machining of nitinol shape memory alloy. *Mater. Des.* **2016**, *109*, 1–9. [CrossRef]
17. Li, C.; Zhang, B.; Li, Y.; Tong, H.; Ding, S.; Wang, Z.; Zhao, L. EDM/ECM high speed drilling of film cooling holes. *J. Mater. Process. Technol.* **2018**, *262*, 95–103. [CrossRef]
18. Dong, T.; Gao, C.; Li, L.; Pei, Y.; Li, S.; Gong, S. Effect of substrate orientations on microstructure evolution and stability for single crystal superalloys in rapid solidification process. *Mater. Des.* **2017**, *128*, 218–230. [CrossRef]
19. Shang, Y.; Zhang, H.; Hou, H.; Ru, Y.; Pei, Y.; Li, S.; Xu, H. High temperature tensile behavior of a thin-walled Ni based single-crystal superalloy with cooling hole: In situ experiment and finite element calculation. *J. Alloys Compd.* **2019**, *782*, 619–631. [CrossRef]
20. Shi, Z.X.; Liu, S.Z.; Yue, X.D.; Wang, Z.C. Effect of Long Term Aging at 980 °C on Microstructure Stability of DD15 Single Crystal Superalloy. *Fail. Anal. Prev.* **2020**, *15*, 217–220.

Disclaimer/Publisher's Note: The statements, opinions and data contained in all publications are solely those of the individual author(s) and contributor(s) and not of MDPI and/or the editor(s). MDPI and/or the editor(s) disclaim responsibility for any injury to people or property resulting from any ideas, methods, instructions or products referred to in the content.

Article

Fatigue-Damage Initiation at Process Introduced Internal Defects in Electron-Beam-Melted Ti-6Al-4V

Robert Fleishel [1,*], William Ferrell [2] and Stephanie TerMaath [1]

[1] Mechanical, Aerospace, and Biomedical Engineering, University of Tennessee, Knoxville, TN 37996, USA
[2] Material Science and Engineering, University of Tennessee, Knoxville, TN 37996, USA
* Correspondence: rfleishe@vols.utk.edu

Abstract: Electron Beam Melting (EBM) is a widespread additive manufacturing technology for metallic-part fabrication; however, final products can contain microstructural defects that reduce fatigue performance. While the effects of gas and keyhole pores are well characterized, other defects, including lack of fusion and smooth facets, warrant additional investigation given their potential to significantly impact fatigue life. Therefore, such defects were intentionally induced into EBM Ti-6Al-4V, a prevalent titanium alloy, to investigate their degradation on stress-controlled fatigue life. The focus offset processing parameter was varied outside of typical manufacturing settings to generate a variety of defect types, and specimens were tested under fatigue loading, followed by surface and microstructure characterization. Fatigue damage primarily initiated at smooth facet sites or sites consisting of un-melted powder due to a lack of fusion, and an increase in both fatigue life and void content with increasing focus offset was noted. This counter-intuitive relationship is attributed to lower focus offsets producing a microstructure more prone to smooth facets, discussed in the literature as being due to lack of fusion or cleavage fracture, and this study indicates that these smooth flaws are most likely a result of lack of fusion.

Keywords: fatigue; titanium; electron beam melting; process defects; damage initiation

1. Introduction

Electron Beam Melting (EBM), also referred to as Selective Electron Beam Melting (SEBM) or electron beam powder bed fusion (E-PBF), is a common powder-bed-based additive manufacturing (AM) technology for the fabrication of metallic parts [1–4]. To manufacture a part, metallic powder is spread over a baseplate. Then an electron beam, generated via a tungsten filament, is used to melt the powder to previously deposited layers or the baseplate in the case of the first layer. EBM is performed in a vacuum, allowing for an elevated chamber temperature higher than that in most selective laser melting (SLM) methods [5–8]. This higher temperature reduces the need for heat treatment to relieve stress in the components [9]. Other advantages of EBM include deep penetration and low reflection into the powder, a high melt rate, low internal stress, energy efficiency, high packing density of parts, and the ability to build parts with no or limited support. In particular, Ti-6Al-4V, a widely used titanium alloy in its wrought form, has seen extensive application in EBM manufacturing.

Depending on process settings and conditions, the EBM process can produce parts containing varying defect types and distributions. It is established that fatigue life in metallic materials can be influenced by such microstructural attributes due to free surface and stress concentration effects [10]. Therefore, the phase transformations and generation of defects during the EBM process for titanium alloys have been previously investigated [5,7,11–14]. These studies identified common defect types, including voids and incomplete melting and/or fusion of powder (lack of fusion). Voids are formed due to trapped gas caused by gas release during the powder melting (gas pores) or trapped gas at the beam tip during high intensity (keyhole pores). An additional defect type, smooth facets [15,16], have been observed on fracture surfaces in multiple studies and are typically attributed to brittle fracture across similarly oriented grain boundaries. Increases in defect generation and porosity were linked to deviations of individual processing parameters from optimal values and decreasing electron beam energy density. Beam energy density is a measure of the beam power acting on a unit of area on the part surface [17]. Beam energy density is thus affected by the beam power, scanning speed, hatch spacing, and beam width.

Fatigue testing of EBM Ti-6Al-4V has been previously conducted through stress-based fatigue testing [7,15,16,18–20], linear elastic fracture mechanics crack-growth-rate testing [11,21], and short crack analysis [18]. The investigations typically focused on optimizing material performance through process-parameter settings. These studies generally noted that the fatigue performance of EBM Ti-6Al-4V is influenced by both the defect population and changes in microstructure due to varying processing conditions. All of the common types of defects previously listed were identified as potential fatigue-damage-initiation sites. Lack of fusion and smooth facet defects were shown to be reduced or eliminated with optimized processing conditions. Therefore, gas and keyhole pores were identified as the primary defects of concern in dictating the material's fatigue performance and were the focus of most studies.

As a result, lack of fusion and smooth-facet defects are not as well characterized, even though their presence may significantly reduce fatigue life. Fatigue specimens containing smooth facet flaws have been found to result in lower fatigue lives, such as outliers in fatigue data sets [16], with large flaws being particularly harmful [14,22,23]. Un-melted powder and smooth facets are probabilistically less likely to result from optimized fabrication than gas or keyhole pores. However, these defects can occur due to processing deviations or aleatoric uncertainty, requiring the inclusion of their effects on fatigue life in damage-tolerance analysis, material qualification, and part certification.

Therefore, the objective of this paper is to intentionally induce these less characterized defects generated by the EBM of Ti-6Al-4V and investigate their effects on fatigue life. While the primary objective is to evaluate the reduction in fatigue life due to these larger and less common defects, the relationship of defect morphology to material failure is also explored. Specimens were printed, using EBM, under three different focus offsets (FOs) that are outside the range of typical printing. The FOs were selected to produce the less common defects of un-melted powder and smooth facets. Stress-controlled fatigue testing was performed to populate an S-N graph. Micrograph characterization and fractography of failed specimens provided information on the microstructural defects and crack-initiation sites.

2. Materials and Methods

2.1. Material Information

Ti-6Al-4V is a classic alpha/beta dual-phase titanium alloy with widespread engineering applications, particularly in aerospace. Extensive characterization of Ti-6Al-4V has been performed in the past decade for process and material development specific to the EBM process [12,15,18,19,21,24–26] and is the material investigated in this study due to its prevalent use.

2.2. Defect Generation

Defect generation in EBM is strongly influenced by the energy density of the electron beam. Energy density can be controlled by changing the focus offset (FO) of the electron beam, while keeping other processing parameters (such as beam power, scanning speed, etc.) unchanged [13]. The FO alters the current through the electromagnet/coil that the electron beam passes through prior to focusing on the powder layer. Changing the amount of current through this coil magnet offsets the focus plane of the electron beam, effectively changing the beam area and consequently the energy density, melt depth, and melt width [27–30]. Therefore, a higher FO produces a wider beam and lower energy density. The relationship between FO and thermal distribution, as well as melt pool size, has been modeled [30], as thermal distribution is an important component in printing quality. The FO parameter in this paper will be referred to in units of electrical current (mA). Note that because the effects of this coil are dependent on the machine configuration, such as the number of wraps and the coil gauge, the effects of FO and range for optimal properties are machine dependent.

A change in FO creates differing types and patterns of defect formation that subsequently result in varying material behavior. Lower FOs lead to a smaller beam area and deeper melt depth, and higher FOs lead to a wider beam area and a shallower melt depth (Figure 1). When the FO is low, the spot size is small, leaving the part susceptible to void formation between electron beam passes, as well as a deeper melt pool, which can impact the previously built layers by re-melting them. This deeper melt pool often results in a keyhole-type weld between the layers. If the FO is high, the spot size becomes large, and the melt-pool overlap is reduced. This wide but shallow melt pool typically results in a conduction-type weld between layers of material. In this case, there is a possibility that the melt plane is above some of the un-melted powder. As a result, high FOs have demonstrated significant increases in un-melted-powder defects [13].

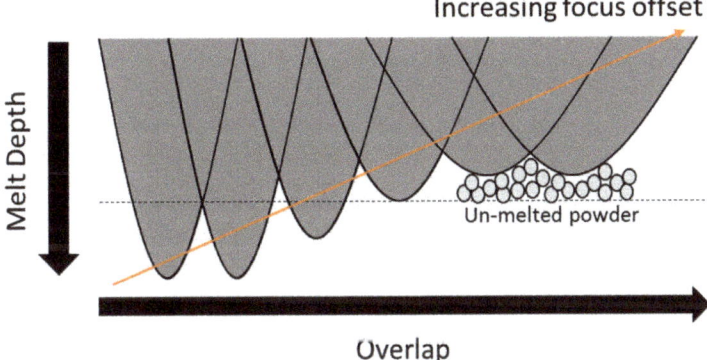

Figure 1. As the FO is increased, the spot shape changes, influencing the melt plane and shape of the melt pool.

2.3. Fatigue Specimen Preparation

EBM was performed using an ARCAM machine with Ti-6Al-4V powder consisting of a 45–120 μm size distribution. Cylindrical volumes of material with a diameter of 15 mm were fabricated with a vertical build orientation, and all processing parameters other than the FO were kept constant for all specimens. These processing parameters, such as beam power, layer thickness, etc., were proprietary to the material supplier and not provided, as the objective was to investigate the effects of the defects, not their cause. Defects were generated in the material by adjusting the FO in the specimen gauge length to create varying defect sizes and shapes based on melt pool overlap and depth. To induce regions of un-melted powder, relatively high FOs were applied. Cylinders were printed using 36 mA, 44 mA, and 52 mA to form the test section area, and a lower FO was used to print the grip sections of cylinders. None of the cylinders were postprocessed to maintain the as-printed microstructure. Nine cylinders were printed for each FO. Each cylinder was machined via lathe to create a constant radius and reduced area gauge length. The reduced section's diameter was 7.5 mm, and the machined specimen's surface finish was left unaltered (Figure 2).

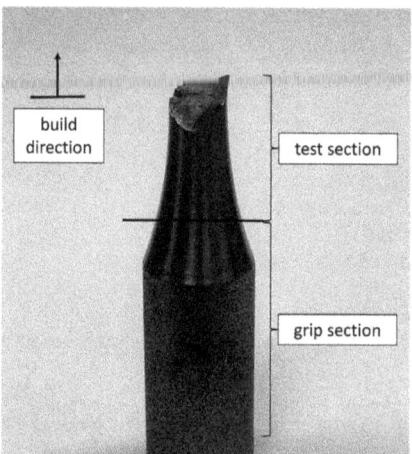

Figure 2. Specimen geometry, with gauge section manufactured using varying FOs.

2.4. Mechanical Testing

To determine maximum stress values for the fatigue testing, an additional set of four specimens produced using the 36 mA FO were tested to tensile failure at a load rate of 0.02 mm/s (corresponding to a strain rate of 4% strain/min) to measure material properties. Average tensile testing results for these four specimens were a Youngs's Modulus of 117.2 GPa, 0.2% offset yield strength of 993 MPa, ultimate tensile strength of 1055 MPa, and an elongation of 8.8% at fracture. The 36 mA EBM specimens in this study are stronger, slightly stiffer, and less ductile than both wrought EBM Ti-6Al-4V and other reported values for EBM Ti-6Al-4V [15].

Stress-controlled fatigue testing was performed using an MTS servo-hydraulic load frame (MTS Systems, Eden Prairie, MN, USA) at a frequency of 10 Hz at room temperature, ambient conditions. ASTM E466 was used in conducting the testing. Specimens were tested to failure, using a load ratio of R = 0.05 at 344.7 MPa, 379.2 MPa, and 413.7 MPa maximum stress levels. Due to a testing error, all specimens printed at 36 mA were tested at 413.7 MPa. The first groups of specimens were tested at 413.7 MPa based on the tensile test results. This value was chosen as slightly less than half of the yield strength. The original testing plan was to then increase this stress for the next group of specimens. Given the failure at relatively low numbers of cycles for all FOs, it was decided to instead lower the maximum stress for the other two testing levels.

2.5. Fracture and Microscopy

Scanning electron microscopy (SEM) (Zeiss Auriga, Oberkochen, Germany) was performed to analyze the fracture surface, identify and view fatigue initiation sites, and characterize crack propagation. The number of damage-initiation sites and large flaws on the fracture surfaces, which were large enough to be visible to the eye, were counted on fracture surfaces, using a 3× magnifying glass. Micrographs were taken by sectioning the specimen below the fracture surface after fatigue testing. Sections were polished with 120 to 1200 grit polishing paper under constant fluid coolant with a final 0.5-micron alumina slurry to obtain the desired polished finish. Micrograph specimen surfaces were etched at room temperature, using a solution of 1% ammonium bifluoride for 60–90 s. These specimens were then analyzed through an SEM or optical microscope (Zeiss Axio Observer -A1, Oberkochen, Germany) to determine the relative porosity to correlate the microstructure characteristics to failure mechanisms.

3. Results

3.1. Micrographs, Microstructure, and Voids

Micrographs for each FO group were taken after etching to observe the microstructure (Figure 3). Microstructures primarily consisting of fine acicular alpha grains [31,32] were observed for all FO groups, indicating a relatively fast cooling rate for all specimens. Regions of high aluminum content (lighter color) and low aluminum content (darker color) were also observed.

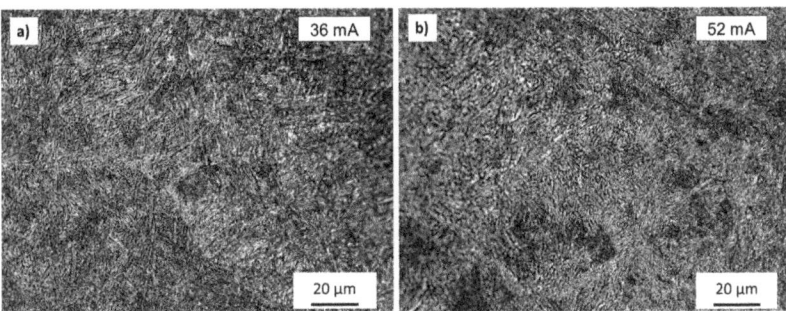

Figure 3. Representative micrographs of material manufactured via different FOs (**a**) 36 mA FO and (**b**) 52mA FO.

SEM imaging of polished sections after fatigue testing showed voids and porosity in all material groups. The voids observed on polished sections were categorized into two groups based on their size: large regions of un-melted powder (Figure 4a) and smaller voids, referred to as microvoids, due to either gas release or lack of fusion between fully melted material (Figure 4b). The size of the regions of un-melted powder were larger, ranging from 100 μm to 450 μm across, compared to the microvoids, which were typically in the range of 2–20 μm across. Lack of fusion voids were present in all material groups. These un-melted powder regions exhibited rough edges due to the un-melted particles and were observed on fracture surfaces across FOs, though their average size and quantity varied. The smooth facets observed on the fatigue-fracture surfaces were not present in the micrographs, though this could be due to the planar nature of the facets being parallel to the direction of sectioning, thus being less likely to appear in the material sections.

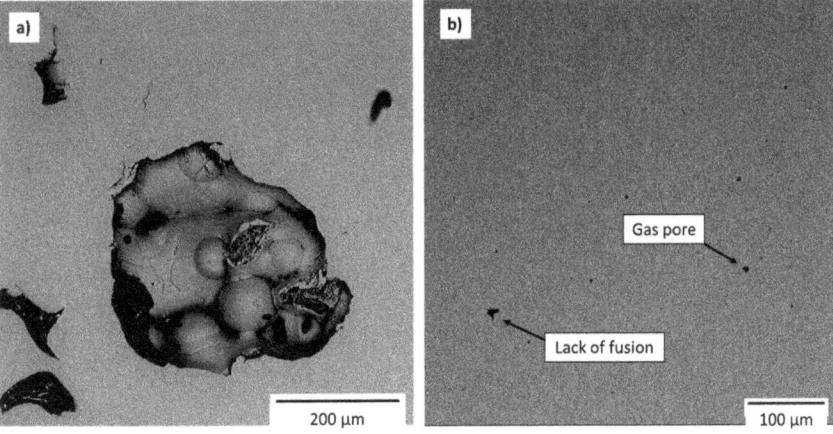

Figure 4. (**a**) Voids caused by un-melted powder and lack of fusion. (**b**) Microvoids caused by gas release and lack of fusion.

The observed microvoids are either the result of trapped argon gas caused by the atomization process or small regions of lack of fusion [33]. The gas pores observed were relatively smooth, with a spherical or elliptical shape, whereas the microvoids attributed to a lack of fusion were more elongated and potentially contained a rough edge similar to those seen in the un-melted powder regions. In a prior study [12], gas pores were observed in EBM Ti-6Al-4V in the range of 5–200 µm, using computed tomography (CT). The gas pores observed in this study were smaller, though this could be attributed to the section imaging used in this study versus the CT method used in the comparison study or the result of different processing parameters. Larger gas pores were not seen in these micrographs, and little evidence of larger gas pores was identifiable on the specimen fracture surfaces.

The microdefects were orders-of-magnitude smaller and much more frequent than the large voids comprising the un-melted powder regions. The frequency of both types of small voids relative to each FO was quantified as the number of voids per mm^2 observed on the micrographs for each specimen. Both the average number of microvoids per unit area and the standard deviation across specimens were calculated for each FO group. The results are provided in Table 1. The total number of microvoids per unit area was quantified, as well as the number attributed to the specific type of microvoid (gas pore or lack of fusion). It should be noted that the number of voids per unit area is provided, not to be confused with a percent porosity measurement.

Table 1. Number of microvoids per mm^2 averaged across all specimens in an FO group. Standard deviation is provided in parentheses.

Focus Offset	Total Microvoids	Gas Pore Voids	Lack of Fusion Voids
36 mA	69.8 (SD: 45.8)	52.3 (SD: 34.3)	17.4 (SD: 21.2)
44 mA	111.9 (SD: 28.8)	79.0 (SD: 30.4)	32.8 (SD: 9.8)
52 mA	102.9 (SD: 40.2)	57.8 (SD: 42.4)	45.2 (SD: 19.8)

SD: standard deviation.

3.2. Fatigue Testing Results

Three specimens in the 44 mA and 52 mA FO groups failed early on in the fatigue test (at 8 and 10 cycles, at 413.7 and 379.2 MPa, respectively for the 44 mA FO specimens and 17 cycles at 379.2 MPa for the 52 mA FO specimen). Failure occurred outside of the gauge length at the interface between the test section material and the grip material. One specimen in the 44 mA group was tested at an incorrect maximum stress due to a machine error. These four specimens were excluded from the analysis.

The valid fatigue testing results exhibited a typical trend for this material, with cycles to failure (N) decreasing with increasing maximum applied stress (S), roughly linearly in maximum stress vs. log(n) (Figure 5). Statistical information for the FO groups at each maximum stress level is provided in Table 2. The average fatigue life increased with FO at the 413.7 MPa and 379.2 MPa maximum stress levels. Identification of a similar trend at the 344.7 MPa maximum stress level is inconclusive due to the limitation of only one valid specimen at the 44 mA FO and none at the 36 mA FO. At the 52 mA FO, two of the specimens failed at nearly identical values of 52,895 and 53,580 cycles to failure, and the third specimen failed at 97,948 cycles. Meanwhile, the 44 mA specimen failed at 75,451 cycles, splitting the difference between the 52 mA FO values. Given the limited number of specimens, there are not enough data points to reasonably conclude a trend at the lowest stress level. These observations are discussed relative to defect type in the discussion section.

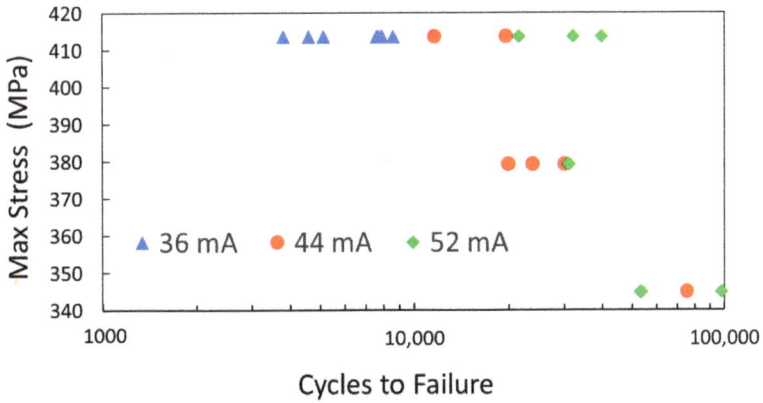

Figure 5. S-N data for the FO groups under high cycle fatigue.

Table 2. Average number of cycles to failure, standard deviation, and number of specimens for varying FO and maximum stress level.

FO (mA)	Maximum Stress Level								
	344.7 MPa			379.2 MPa			413.7 MPa		
	Avg	SD	#	Avg	SD	#	Avg	SD	#
36	–	–	–	–	–	–	6484	1800	9
44	75,451	23,049	1	24,628	5080	3	15,660	5681	2
52	68,141	25,816	3	30,994	186	2	31,181	9134	3

Avg, average; SD, standard deviation; #, number of specimens.

3.3. Fracture Surface Defect Classification

The imaging of the fatigue fracture surfaces showed similar types of porosity and defects on the fracture surfaces of all three FO groups, though there was an observed difference in the number of potential damage-initiation sites between FO groups. Defects were characterized into three categories, with examples of each displayed in Figure 6. Several sites contained multiple defect types/classifications. Representative examples are presented in the discussion section.

1. Deep un-melted powder (Figure 6a): Defects due to the incomplete melt of powder and lack of fusion, where remnants of spherical powder are clearly observed.
2. Smooth facets (Figure 6b,c): Smooth flaws that lack clear spherical powder remnants, appearing as outlines, potentially due to semi-fused powder or some microstructural feature, such as grain boundaries, are present. An additional texture of parallel lines running across the flaws is also observed on some facets (Figure 6c). Higher magnification of the surface texture on these flaws is shown in Figure 7.
3. Gas pores (Figure 6d): Round flaw formed by gas release during the melt process.

While defects on the fracture surfaces were relatively easy to categorize based on visible features, classifying and describing their formation is more difficult. Un-melted powder (Figure 6a) and gas-pore defects (Figure 6d) can easily be attributed to an incomplete melting of powder and lack of fusion and gas release, respectively. Both defect types are well documented in the EBM Ti-6Al-4V literature previously discussed. Fracture surface defects similar to the smooth flaws in this study, both with surface lines [16,19,34] and without surface lines [15,34], have been observed in other studies of fatigue of EBM Ti-6Al-4V. The cause of these type of defects has been attributed to multiple mechanisms, primarily either from lack of fusion [19,34] or due to cleavage fracture [16,35]. Cleavage fracture is suggested to occur by quasi-cleavage along grains or similarly oriented grain colony boundaries loaded perpendicularly to their basal plains. The observed smooth facets occurred in a variety of locations within the specimens, both near the surface and close to the center in a variety of shapes. These smooth facets created a shiny region in the fracture surface. Smooth facets were present on the vast majority of specimen fracture surfaces, often in multiple locations, and were most prevalent in the 36 mA set.

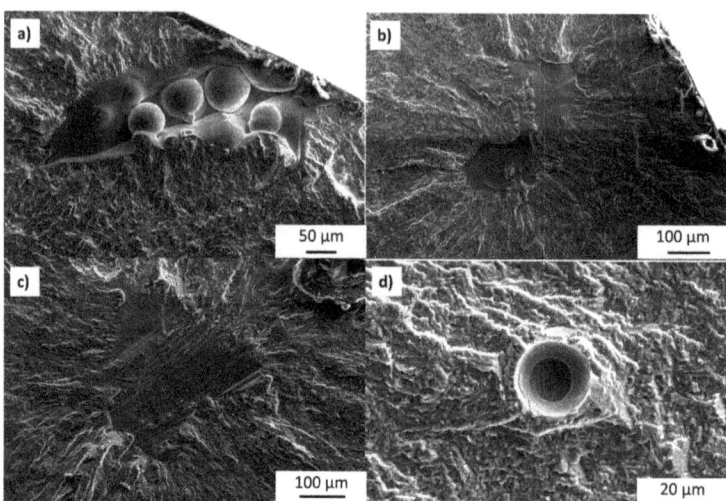

Figure 6. Fracture-surface images of observed defect types. (**a**) Deep un-melted powder. (**b**) Smooth facet without line texture. (**c**) Smooth flaw containing a line texture. (**d**) Gas pore.

Figure 7. Texture of smooth facets (**a**) without lines and (**b**) with surface lines.

3.4. Fracture Surface Observations and Trends

Failure generally appeared to initiate at regions where un-melted powder and/or smooth facets were present. Some evidence of gas pores was also present on the fracture surfaces, but these flaws were significantly smaller and less common than those due to un-melted powder or smooth defect failure and deemed less impactful to fatigue failure for the material used in this study. The average number of defects observed with a 3× magnifying glass and the average largest flaw size in each FO are provided in Table 3. Histograms displaying the distribution of the number of flaws and largest flaw diameter observed on each specimen's fracture surface are shown in Figure 8. While no significant difference is discernible between the 44 mA and 52 mA groups in terms of number of flaws, by comparison, the lower 36 mA group is distinguished by a substantially higher number of flaws per fracture surface. Other than the two outliers, there is no notable statistical difference for the size of the largest flaws on fracture surfaces across all groups. The specimen numbers for the outliers are identified on Figure 8b for further discussion.

Table 3. Summary of fracture surface measurements for each FO group.

FO	Average Largest Flaw Size (μm)	Average Number of Flaws on Surface
36 mA	577 (SD: 198)	26.0 (SD: 4.1)
44 mA	578 (SD: 322)	10.8 (SD: 2.1)
52 mA	537 (SD: 302)	10.9 (SD: 3.4)

SD: standard deviation.

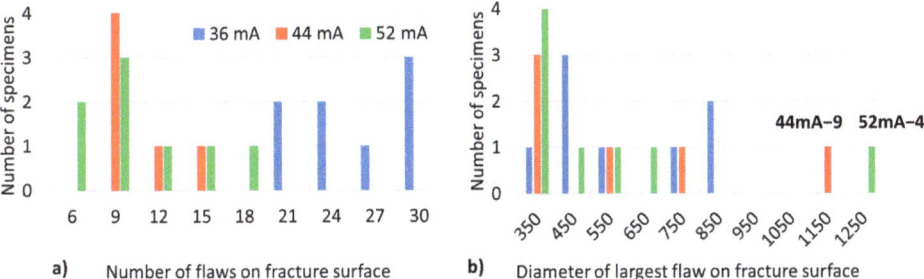

a) Number of flaws on fracture surface
b) Diameter of largest flaw on fracture surface

Figure 8. Histogram of (a) the number of fatigue flaws present on each specimen's fracture surface and (b) the size of the largest flaw present on each specimen's fracture surface.

4. Discussion

4.1. General Trends

The most easily observable and initially surprising trend observed in the fatigue-testing data is that fatigue life appeared to increase with increasing FO, contrary to a previous study [13], wherein the higher energy density of lower FOs was shown to exhibit a lower defect formation. Two potential explanations for this difference are the number of optically observable flaws present (Figure 9a) and the type of flaws on the fracture surfaces (Table 4). For the relationship between the number of initiating flaws on the fracture surface versus fatigue life, a general negative relationship is clear. This correlation was also established by Tammas-Williams [16], who concluded that close proximity between defects could increase stress concentrations, thus increasing the potential for crack nucleation. A greater number of defects increases the probability of defects in close proximity. However, it is difficult to claim that the number of flaws was the primary factor in fatigue resistance because of the lack of a trend within individual FO groups. This is most clearly seen in the 36 mA data, where there is minimal variation in fatigue life even with a large change in the number of fatigue-initiating flaws. Additionally, the fatigue life is separated by FO groups. However, this observation may be due to other interacting factors, such as the flaw type and location.

Table 4. Percent of specimens within each group showing signs of each type of flaw.

FO Group	Deep Un-Melted Powder	Smooth Facets without Line Texture	Smooth Facets with Line Texture	Gas Pore
36 mA	25%	100%	75%	37%
44 mA	17%	100%	33%	17%
52 mA	25%	100%	43%	37%

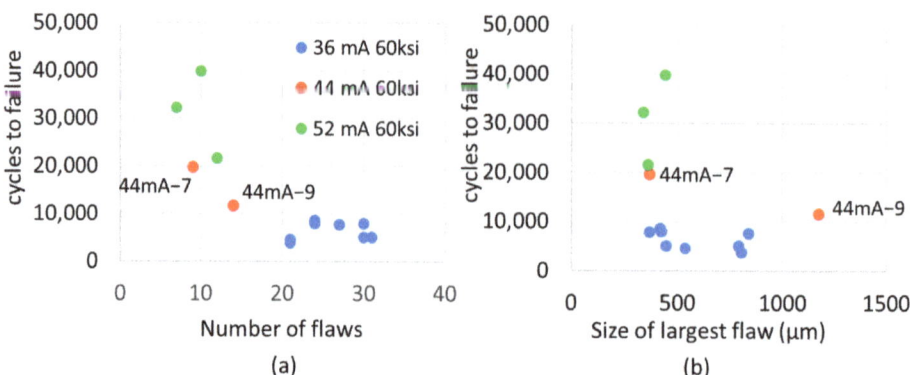

Figure 9. Plots comparing fatigue life at 413.7 MPa max stress level to (**a**) the number of flaws observed on fracture surfaces and (**b**) the size of the largest flaw present on each fracture surface.

Therefore, in addition to the number and size of flaws present on the fracture surfaces, we investigated trends in the types of defects present on the fracture surfaces of each group to explain the difference in fatigue lives. The prevalence of different types of flaw classifications in each FO group is shown in Table 4. Evidence of each type of flaw is present in each group, with relatively uniform occurrence across all FO groups, except for smooth facets with surface lines, which were much more prevalent in the 36 mA group than in the 44 mA or 52 mA groups, potentially contributing to the lower fatigue life of the 36 mA group. A prior study by Rafi [15] also noted such smooth facets at fracture-initiation sites. There is, however, limited evidence in the statistics to explain the difference in fatigue lives between the 44 mA and 52 mA sets.

To explore this difference, the fracture surfaces for the 44 mA specimens tested at 413.7 MPa that exhibited lower fatigue lives than their 52 mA counterparts (Figure 9) are provided in Figure 10. Both specimens contained smooth facet flaws with surface lines, and these flaws are on or near the specimen surfaces. It is well established in the literature that pores near the surface shorten the fatigue life of specimens produced by both EBM and SLM [16,36–40]. Moreover, Xu et. al. [41] demonstrated the increase in stress concentration caused by pores within one diameter of the specimen surface relative to internal pores through a finite element study. It is evident from these specimens that these conclusions also pertain to smooth facet flaws. Both factors (flaw type and location) could account for the shortened fatigue life of the 44 mA specimens relative to the 52 mA specimens.

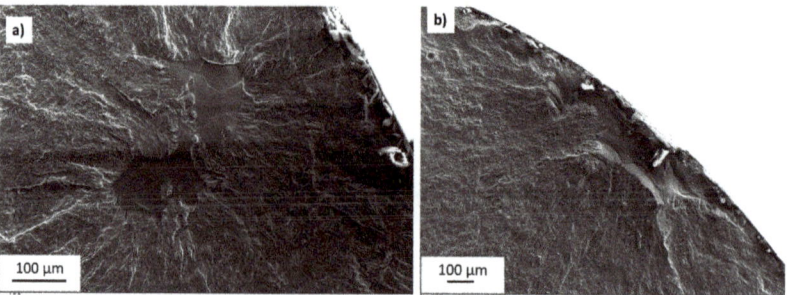

Figure 10. Micrographs of 44 mA specimens tested at 413.7: (**a**) 44mA−7 and (**b**) 44mA−9.

Alternatively, when comparing the fracture surfaces of the 44 mA and 52 mA FO specimens at 344.7 MPa, no significant differences were noted. The number, type, and size of defects were statistically consistent across the specimens of both FO groups. It can therefore be reasonably hypothesized that the difference in fatigue lives among specimens and lack of a trend between groups at this stress level is due to aleatoric scatter. This finding is considered a hypothesis and not a conclusion, as there was only one valid specimen in the 44 mA group. Therefore, there are not enough data points to conclusively state that there are no other material differences between the groups that could cause a difference in fatigue life that would be noticeable with more data.

There is no conclusive trend relative to the size of the largest flaw vs. fatigue life in general (Figure 9b). Specimens containing flaws of approximately the same size exhibited substantial variance in fatigue life. When differences did exist between specimens with significantly different flaw sizes, it was not discernible whether flaw size alone dictated the variation in fatigue life. For example, specimen 44mA−9 contained a significantly larger surface flaw than the other 44 mA specimen tested at the same stress level of 413.7 MPa. Specimen 44mA−9 failed at 11,643 cycles relative to specimen 44mA−7, which failed at 19,677 cycles. As discussed in the previous paragraph, both specimens contained a smooth facet with surface lines; however, the smooth facet observed in specimen 44mA−9 was closer to the surface than specimen 44mA−7. There is not enough data to conclusively determine if the difference in fatigue life is due to the flaw being closer to the surface, the flaw being considerably larger, or aleatoric uncertainty. The other outlier, 52mA−4, was tested at 379.2 MPa and failed at nearly the same number of cycles (30,862) as its counterpart specimen 52mA−2 (31,125). Both specimens exhibited similar fracture surfaces with multiple flaws and smooth facet flaws, which seem to be more influential than flaw size. Prior studies have concluded that the flaw area normal to loading is more dominant in shortening fatigue life than the total volume of the flaw [40,42]. From the presented results and prior studies, it appears that flaw type, location, and other factors are more influential than the size of the flaw.

A trend that is notably lacking is that flaws containing deep un-melted powder are not more likely to be present on fracture surfaces of the higher (44 mA and 52 mA) FO groups. This finding is surprising, because regions of un-melted powder have been reported to serve as fatigue-initiation sites and to be more likely to occur in high-FO conditions [15]. Additionally, the prevalence of microvoids due to the lack of fusion was observed at higher FOs. Based on the specimen fracture surfaces, the smaller pores do not seem to influence the cycles to failure of the high-FO groups for the investigated fatigue loadings. Due to the relatively small size and the spherically dominated shape of the pores caused by gas release, these pores may be significant in more dense prints and may pose problems when evaluating the density of the parts [43].

4.2. Classification of Defects

Further interpretation of the results strongly depends on classification of the defects observed on the fracture surface, especially of the smooth facets (Figure 6b,c), due to their prevalence in the 36 mA group, which exhibited the shortest fatigue life. Two types of lack-of-fusion flaws were observed: (1) defects with observable un-melted powder and (2) fully melted material in layers above and below the facet, resulting in the smooth texture, but with no or weak bonding between layers. As previously mentioned, the two competing explanations for the smooth facets (both with and without a line texture) are either from lack of fusion [14,34] or cleavage fracture [16,35]. An interesting result from the current study is that the highest FO (52 mA) specimens exhibited better fusion and less porosity than those printed at the lowest FO value (36 mA), going against the literature trends [12,15,29] or possibly suggesting that there is an FO value above which the lack of fusion begins to decrease or is less detrimental to fatigue behavior. If the smooth facets are caused by a lack of fusion, the surface lines would be attributed to solidification lines forming as the material cools after melting. This explanation, however, does not explain the higher incidence of surface lines on the smooth facets in the 36 mA group relative to the higher-FO groups. Alternatively, if the smooth facets are attributed to a quasi-cleavage fracture, then the quasi-cleavage of grains is a more important factor in the fatigue resistance of EBM Ti-6Al-4V than void content due to lack of fusion for this set of manufacturing parameters.

This could be due to the fact that the FO and other processing parameters used in this study are more likely to produce neighborhoods of grains with the basal plain perpendicular to the applied load. In this case, the surface lines seen on some of the smooth defects would be attributed to steps or changes in the cleavage fracture plane.

A further explanation of the cause of these defects is provided in Figure 11, showing magnified images of some regions of interest near smooth flaws. These images show features of both smooth texture with several un-melted powder beads in and around the flaw (Figure 11a,c). The presence of un-melted particles retaining their spherical shape on the surface is evidence of a lack of fusion. Figure 11d also supports this conclusion, though instead of spherical powder particles, the smooth facet texture (shown with surface lines) is interspersed with regions of ductile fracture. These regions of ductile fracture, with a texture suggesting void coalescence, could be due to particles that partially fused across the smooth flaw. The image in Figure 11b shows a higher magnification of surface texture on smooth facets, revealing a texture similar to the fine acicular alpha microstructure seen in the micrographs. This could be the texture above the melt if the smooth flaw was due to lack of fusion, or it could be the cleavage fracture through the microstructure, pointing to the quasi-cleavage mechanism for the formation of the smooth facet flaws.

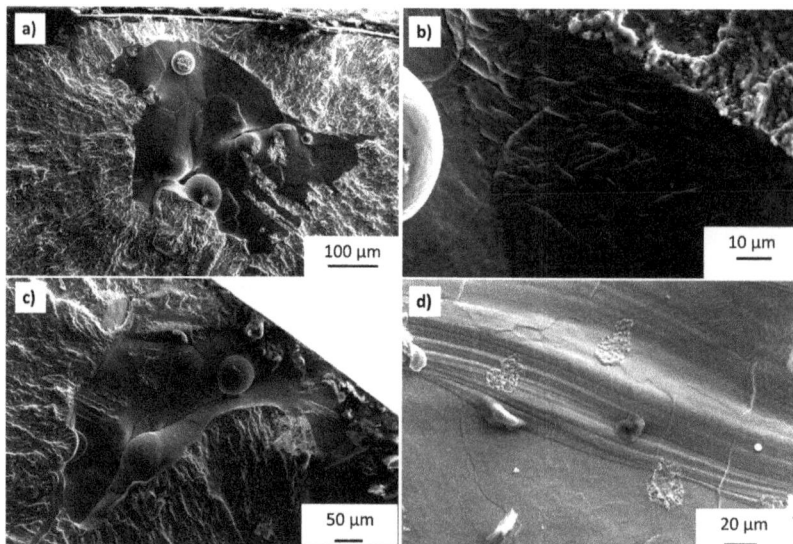

Figure 11. (**a**) Smooth flaw, no surface lines, and what appears to be a near-fully un-melted powder particle on the flaw surface. (**b**) Closer image of the flaw shown in (**a**), with a texture similar to the fine acicular alpha microstructure seen in the micrographs. (**c**) Image of smooth flaw, where the distinction between un-melted particles and microstructural outlines is blurred. (**d**) Close image of a smooth flaw with surface lines, with areas containing a texture indicative of ductile fracture.

Overall, the images present more evidence for a lack of fusion as the source of these flaws, though we cannot completely rule out a cleavage fracture mechanism. Care should be taken in classifying defects of this type in future studies due to the multiple possible explanations. An interesting consequence of this conclusion is the different qualities of these two types of lack of fusion defects; the flaws where beads of powder remain un-melted (described as deep un-melted powder elsewhere in this paper) and the flaws where the material fails to fuse across the melt plane (the smooth flaws). This distinction could explain the increase in the smooth lack of fusion defects and shorter fatigue lives as FO decreases in this study, with the smoother flaws containing less free space while still weakening the material.

4.3. Comparison to Other Data

When the fatigue data collected in this study are compared to fatigue data published in other research on EBM Ti-6Al-4V [15,16] (Figure 12), the fatigue life measured in this study consistently underperforms. The fatigue data in these other studies were collected using R = 0.1 and R = 0.0, respectively, and the results were normalized to the R = 0.05 testing value in this study, using the Smith–Watson–Topper (SWT) mean stress correction [44]. A critical observation is the extreme drop in fatigue life relative to the Tammas-Williams [16] data, even though the specimens in the current study were tested at significantly lower stress levels. The current data also exhibited significantly lower fatigue life than the Rafi data [15], which were obtained at similar stress levels.

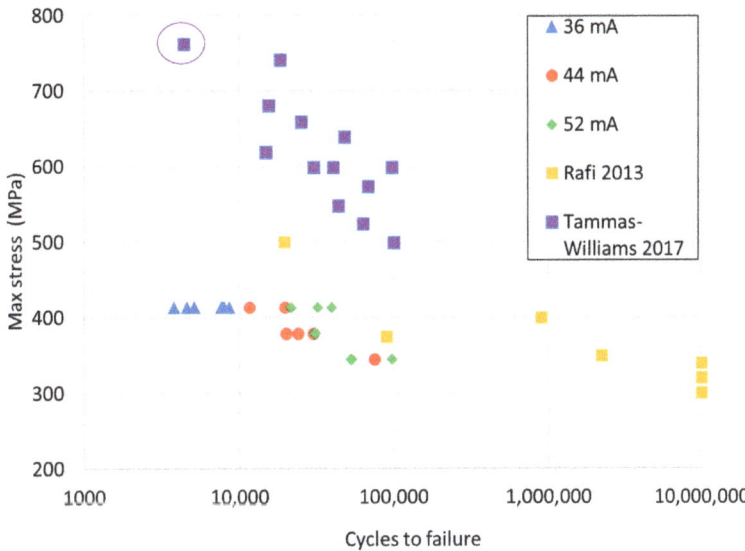

Figure 12. Fatigue-test data comparison to published results. An early failure in the Tammas-Williams data is highlighted, as it contained a smooth facet flaw on the fracture surface.

The higher prevalence of the smooth facets and the lower prevalence of keyhole pores in this study versus those seen in the Rafi [15] and Tammas-Williams [16] specimens provided a higher number of large potential fatigue-damage-initiation sites. This difference in flaw types may explain the poorer fatigue behavior of the 36 mA FO group compared to the 44 mA and 52 mA groups, due to larger number of smooth facet sites in the 36 mA FO group than the others in this study. It is interesting to note that the circled data point in the Tammas-William data was identified as fracturing at a smooth facet, and this specimen failed earlier than the others. This specimen was also tested at a slightly higher maximum stress than the next stress level; however, the specimen failed at a significantly lower number of cycles. Additional factors, such as a more conduction-type weld in this study versus a keyhole-type weld in the others and the microstructure indicating a fast cooling rate in this study are also likely to have negatively impacted the fatigue performance of the material. These data reinforce the extent to which large regions of lack of fusion can degrade fatigue performance, so extra care should be taken in process control and inspection procedures when fatigue damage is a concern.

It is also important to consider the maximum stress levels relative to the yield strength when evaluating the effects of smooth facets on fatigue life. The maximum stress levels as a percentage of yield strength in the current study are 413.7 MPa (42%), 379.2 MPa (38%), and 344.7 MPa (35%). The effects of smooth facets are concerning when considering that all specimens were tested at less than half of the yield strength. This finding has significant implications when determining safety factors.

5. Conclusions

The increased FOs beyond typical values generated voids indicative of insufficient energy density to fully melt the beads, resulting in a substantial lack of fusion flaws. Major takeaways from the fatigue test results and imaging are as follows:

- Failure initiation sites consistently showed signs of smooth facets and/or lack of fusion. Lower FOs led to more smooth facet initiation sites. Damage initiating at these locations had a strong negative effect on the fatigue performance compared to other studies in which these defects were better controlled.
- The higher FOs showed increases in the number of voids and lack of fusion with un-melted powder, but a lower number of smooth facets were present on fracture surfaces, resulting in an increased fatigue life.
- The number of smooth/facet flaws were determined to be the primary factor in differentiating fatigue life performance. These smooth facets are more likely due to a lack of fusion between layers during the EBM process than cleavage fracture, but similar examples of this defect type have been attributed to both mechanisms in the literature, and care should be taken in classifying them in the future.

Author Contributions: Conceptualization, R.F., W.F. and S.T.; methodology, R.F., W.F. and S.T.; formal analysis, R.F.; investigation, R.F., W.F. and S.T.; resources, R.F., W.F. and S.T.; data curation, R.F.; writing—original draft preparation, R.F., W.F. and S.T; writing—review and editing, R.F., W.F. and S.T; visualization, R.F.; supervision, S.T.; project administration, S.T.; funding acquisition, S.T. All authors have read and agreed to the published version of the manuscript.

Funding: This research was funded by the Manufacturing and Materials Joining Innovation Center (Ma2JIC), made possible through awards NSF IIP-1540000 (Phase I) and NSF IIP-1822186 (Phase II) from the National Science Foundation Industry University Cooperative Research Center program (IUCRC). The APC was funded by the University of Tennessee.

Institutional Review Board Statement: Not applicable.

Informed Consent Statement: Not applicable.

Data Availability Statement: Data sharing not applicable.

Conflicts of Interest: The authors declare no conflict of interest. The funders had no role in the design of the study; in the collection, analyses, or interpretation of data; in the writing of the manuscript; or in the decision to publish the results.

References

1. Ataee, A.; Li, Y.; Fraser, D.; Song, G.; Wen, C. Anisotropic Ti-6Al-4V gyroid scaffolds manufactured by electron beam melting (EBM) for bone implant applications. *Mater. Des.* **2018**, *137*, 345–354. [CrossRef]
2. Nicoletto, G.; Konečná, R.; Frkáň, M.; Riva, E. Surface roughness and directional fatigue behavior of as-built EBM and DMLS Ti6Al4V. *Int. J. Fatigue* **2018**, *116*, 140–148. [CrossRef]
3. Koptioug, A.; Rännar, L.E.; Bäckström, M.; Shen, Z.J. New metallurgy of additive manufacturing in metal: Experiences from the material and process development with electron beam melting technology (EBM). In *Materials Science Forum*; Trans Tech Publications Ltd.: Bäch, Switzerland, 2017.
4. Sidambe, A.T. Three dimensional surface topography characterization of the electron beam melted Ti6Al4V. *Met. Powder Rep.* **2017**, *72*, 200–205. [CrossRef]
5. Liu, Y.J.; Li, S.J.; Wang, H.L.; Hou, W.T.; Hao, Y.L.; Yang, R.; Sercombe, T.B.; Zhang, L.C. Microstructure, defects and mechanical behavior of beta-type titanium porous structures manufactured by electron beam melting and selective laser melting. *Acta Mater.* **2016**, *113*, 56–67. [CrossRef]
6. Snyder, J.C.; Stimpson, C.K.; Thole, K.A.; Mongillo, D.J. Build Direction Effects on Microchannel Tolerance and Surface Roughness. *J. Mech. Des. Trans. ASME* **2015**, *137*, 111411. [CrossRef]
7. Zhao, X.; Li, S.; Zhang, M.; Liu, Y.; Sercombe, T.B.; Wang, S.; Hao, Y.; Yang, R.; Murr, L.E. Comparison of the microstructures and mechanical properties of Ti-6Al-4V fabricated by selective laser melting and electron beam melting. *Mater. Des.* **2016**, *95*, 21–31. [CrossRef]

8. Zhang, L.C.; Liu, Y.; Li, S.; Hao, Y. *Additive Manufacturing of Titanium Alloys by Electron Beam Melting: A Review*; Wiley-VCH Verlag: Hoboken, NJ, USA, 2018; Volume 20. [CrossRef]
9. GE Additive. Inside Electron Beam Melting. White paper. 2019. Available online: https://go.additive.ge.com/rs/706-JIU-273/images/GE%20Additive_EBM_White%20paper_v3.pdf.
10. Suresh, S. *Fatigue of Materials*, 2nd ed.; Cambridge University Press: Cambridge, UK, 1998.
11. Seifi, M.; Salem, A.; Satko, D.; Shaffer, J.; Lewandowski, J.J. Defect distribution and microstructure heterogeneity effects on fracture resistance and fatigue behavior of EBM Ti–6Al–4V. *Int. J. Fatigue* **2017**, *94*, 263–287. [CrossRef]
12. Tammas-Williams, S.; Zhao, H.; Léonard, F.; Derguti, F.; Todd, I.; Prangnell, P.B. XCT analysis of the influence of melt strategies on defect population in Ti-6Al-4V components manufactured by Selective Electron Beam Melting. *Mater. Charact.* **2015**, *102*, 47–61. [CrossRef]
13. Gong, H.; Rafi, H.K.; Rafi, K.; Starr, T.; Stucker, B. The Effects of Processing Parameters on Defect Regularity in Ti-6Al-4V Parts Fabricated By Selective Laser Melting and Electron Beam Melting. In Proceedings of the 2013 International Solid Freeform Fabrication Symposium, Austin, TX, USA, 12–14 August 2013.
14. Sanaei, N.; Fatemi, A. Defects in additive manufactured metals and their effect on fatigue performance: A state-of-the-art review. *Prog. Mater. Sci.* **2021**, *117*, 100724. [CrossRef]
15. Rafi, H.K.; Karthik, N.V.; Gong, H.; Starr, T.L.; Stucker, B.E. Microstructures and mechanical properties of Ti6Al4V parts fabricated by selective laser melting and electron beam melting. *J. Mater. Eng. Perform.* **2013**, *22*, 3872–3883. [CrossRef]
16. Tammas-Williams, S.; Withers, P.J.; Todd, I.; Prangnell, P.B. The Influence of Porosity on Fatigue Crack Initiation in Additively Manufactured Titanium Components. *Sci. Rep.* **2017**, *7*, 7308. [CrossRef] [PubMed]
17. Liu, Z.Y.; Li, C.; Fang, X.Y.; Guo, Y.B. Energy Consumption in Additive Manufacturing of Metal Parts. *Procedia Manuf.* **2018**, *26*, 834–845. [CrossRef]
18. Beretta, S.; Romano, S. A comparison of fatigue strength sensitivity to defects for materials manufactured by AM or traditional processes. *Int. J. Fatigue* **2017**, *94*, 178–191. [CrossRef]
19. Hrabe, N.; Gnäupel-Herold, T.; Quinn, T. Fatigue properties of a titanium alloy (Ti–6Al–4V) fabricated via electron beam melting (EBM): Effects of internal defects and residual stress. *Int. J. Fatigue* **2017**, *94*, 202–210. [CrossRef]
20. Chern, A.H.; Nandwana, P.; Yuan, T.; Kirka, M.M.; Dehoff, R.R.; Liaw, P.K.; Duty, C.E. A review on the fatigue behavior of Ti-6Al-4V fabricated by electron beam melting additive manufacturing. *Int. J. Fatigue* **2019**, *119*, 173–184. [CrossRef]
21. Zhai, Y.; Galarraga, H.; Lados, D.A. Microstructure Evolution, Tensile Properties, and Fatigue Damage Mechanisms in Ti-6Al-4V Alloys Fabricated by Two Additive Manufacturing Techniques. *Procedia Eng.* **2015**, *114*, 658–666. [CrossRef]
22. Hu, Y.N.; Wu, S.C.; Withers, P.J.; Zhang, J.; Bao, H.Y.X.; Fu, Y.N.; Kang, G.Z. The effect of manufacturing defects on the fatigue life of selective laser melted Ti-6Al-4V structures. *Mater. Des.* **2020**, *192*, 108708. [CrossRef]
23. Günther, J.; Krewerth, D.; Lippmann, T.; Leuders, S.; Tröster, T.; Weidner, A.; Biermann, H.; Niendorf, T. Fatigue life of additively manufactured Ti–6Al–4V in the very high cycle fatigue regime. *Int. J. Fatigue* **2017**, *94*, 236–245. [CrossRef]
24. Hrabe, N.; Quinn, T. Effects of processing on microstructure and mechanical properties of a titanium alloy (Ti-6Al-4V) fabricated using electron beam melting (EBM), part 1: Distance from build plate and part size. *Mater. Sci. Eng. A* **2013**, *573*, 264–270. [CrossRef]
25. Hrabe, N.; Quinn, T. Effects of processing on microstructure and mechanical properties of a titanium alloy (Ti-6Al-4V) fabricated using electron beam melting (EBM), Part 2: Energy input, orientation, and location. *Mater. Sci. Eng. A* **2013**, *573*, 271–277. [CrossRef]
26. Sidambe, A.T.; Todd, I.; Hatton, P.V. Effects of build orientation induced surface modifications on the in vitro biocompatibility of electron beam melted Ti6Al4V. *Powder Metall.* **2016**, *59*, 57–65. [CrossRef]
27. Schwerdtfeger, J.; Singer, R.F.; Körner, C. In situ flaw detection by IR-imaging during electron beam melting. *Rapid Prototyp. J.* **2012**, *18*, 259–263. [CrossRef]
28. Biamino, S.; Penna, A.; Ackelid, U.; Sabbadini, S.; Tassa, O.; Fino, P.; Pavese, M.; Gennaro, P.; Badini, C. Electron beam melting of Ti-48Al-2Cr-2Nb alloy: Microstructure and mechanical properties investigation. *Intermetallics* **2011**, *19*, 776–781. [CrossRef]
29. Lee, H.J.; Kim, H.K.; Hong, H.U.; Lee, B.S. Influence of the focus offset on the defects, microstructure, and mechanical properties of an Inconel 718 superalloy fabricated by electron beam additive manufacturing. *J. Alloys Compd.* **2019**, *781*, 842–856. [CrossRef]
30. Galati, M.; Snis, A.; Iuliano, L. Experimental validation of a numerical thermal model of the EBM process for Ti6Al4V. *Comput. Math. Appl.* **2019**, *78*, 2417–2427. [CrossRef]
31. Gammon, L.M.; Briggs, R.D.; Packard, J.M.; Batson, K.W.; Boyer, R.; Domby, C.W. Metallography and Microstructures of Titanium and Its Alloys. *ASM Handb.* **2004**, *9*, 899–917. [CrossRef]
32. Li, Y.; Song, L.; Xie, P.; Cheng, M.; Xiao, H. Enhancing hardness and wear performance of laser additive manufactured Ti6Al4V alloy through achieving ultrafine microstructure. *Materials* **2020**, *13*, 1210. [CrossRef]
33. Romano, S.; Abel, A.; Gumpinger, J.; Brandão, A.D.; Beretta, S. Quality control of AlSi10Mg produced by SLM: Metallography versus CT scans for critical defect size assessment. *Addit. Manuf.* **2019**, *28*, 394–405. [CrossRef]
34. Sandell, V.; Hansson, T.; Roychowdhury, S.; Månsson, T.; Delin, M.; Åkerfeldt, P.; Antti, M.L. Defects in electron beam melted Ti-6Al-4V: Fatigue life prediction using experimental data and extreme value statistics. *Materials* **2021**, *14*, 640. [CrossRef]
35. Nalla, R.K.; Boyce, B.L.; Campbell, J.P.; Peters, J.O.; Ritchie, R.O. Influence of Microstructure on High-Cycle Fatigue of Ti-6Al-4V: Bimodal vs. Lamellar Structures. *Metall. Mater. Trans. A* **2002**, *33*, 899–918. [CrossRef]

36. Edwards, P.; O'conner, A.; Ramulu, M. Electron beam additive manufacturing of titanium components: Properties and performance. *J. Manuf. Sci. Eng.* **2013**, *135*, 061016. [CrossRef]
37. Antonysamy, A.A. *Microstructure, Texture and Mechanical Property Evolution during Additive Manufacturing of Ti6Al4V Alloy for Aerospace Applications*; The University of Manchester: Manchester, UK, 2012.
38. Liu, Q.C.; Elambasseril, J.; Sun, S.J.; Leary, M.; Brandt, M.; Sharp, P.K. The effect of manufacturing defects on the fatigue behaviour of Ti-6Al-4V specimens fabricated using selective laser melting. In *Advanced Materials Research*; Trans Tech Publications Ltd.: Bäch, Switzerland, 2014; pp. 1519–1524.
39. Amsterdam, E.; Kool, G. High cycle fatigue of laser beam deposited Ti-6Al-4V and Inconel 718. In *ICAF 2009, Bridging the Gap between Theory and Operational Practice*; Springer: Berlin/Heidelberg, Germany, 2009; pp. 1261–1274.
40. Murakami, Y. Material defects as the basis of fatigue design. *Int. J. Fatigue* **2012**, *41*, 2–10. [CrossRef]
41. Xu, Z.; Wen, W.; Zhai, T. Effects of pore position in depth on stress/strain concentration and fatigue crack initiation. *Metall. Mater. Trans. A* **2012**, *43*, 2763–2770. [CrossRef]
42. Li, P.; Lee, P.; Maijer, D.; Lindley, T. Quantification of the interaction within defect populations on fatigue behavior in an aluminum alloy. *Acta Mater.* **2009**, *57*, 3539–3548. [CrossRef]
43. Chisena, R.S.; Engstrom, S.M.; Shih, A.J. Computed tomography evaluation of the porosity and fiber orientation in a short carbon fiber material extrusion filament and part. *Addit. Manuf.* **2020**, *34*, 101189. [CrossRef]
44. Dowling, N. *Mechanical Behavior of Materials*, 4th ed.; Pearson: London, UK, 2013.

Disclaimer/Publisher's Note: The statements, opinions and data contained in all publications are solely those of the individual author(s) and contributor(s) and not of MDPI and/or the editor(s). MDPI and/or the editor(s) disclaim responsibility for any injury to people or property resulting from any ideas, methods, instructions or products referred to in the content.

Article

Cluster Hardening Effects on Twinning in Mg-Zn-Ca Alloys

Ruixue Liu [1], Jie Wang [1], Leyun Wang [1,2,*], Xiaoqin Zeng [1,2,*] and Zhaohui Jin [3]

[1] National Engineering Research Center of Light Alloy Net Forming, Shanghai Jiao Tong University, Shanghai 200240, China; liurx1221@163.com (R.L.); 18217278361@163.com (J.W.)
[2] State Key Laboratory of Metal Matrix Composites, Shanghai Jiao Tong University, Shanghai 200240, China
[3] Shenyang National Laboratory for Materials Science, Institute of Metal Research, Chinese Academy of Sciences, 72 Wenhua Road, Shenyang 110016, China; zhjin@imr.ac.cn
* Correspondence: leyunwang@sjtu.edu.cn (L.W.); xqzeng@sjtu.edu.cn (X.Z.)

Abstract: Twinning is a critical deformation mode in Mg alloys. Understanding deformation twinning (DT) is essential to improving mechanical properties of Mg alloys. To address the experimentally observed conspicuous hardening effects in Mg-1.8Zn-0.2Ca alloys, interactions between the $\{10\bar{1}2\}$ twin boundaries (TBs) and solute clusters in Mg-Zn-Ca alloys were examined via molecular dynamics (MD) simulations. We find that the Zn/Ca-containing clusters show different hindering effects on TBs and an increment in the applied shear stress of 100 MPa is required to accomplish the interaction between the boundary and the cluster with Ca content > 50 at%. The cluster hardening effects on twinning are positively correlated to the Ca content and the size of the clusters in Mg-Zn-Ca alloys.

Keywords: Mg-Zn-Ca alloy; twin boundary; solute cluster hardening; molecular dynamics simulation

1. Introduction

Lightweight magnesium (Mg) alloys are in the spotlight for energy efficiency for transportation applications [1,2]. Deformation twinning (DT) on the $\{10\bar{1}2\}\langle\bar{1}011\rangle$ system is an important plastic deformation mechanism in Mg with a hexagonal crystal structure. Nucleation and growth of twins are responsible for hardening and texture evolution characteristics, and ultimately influence the mechanical properties and formability of Mg and its alloys [3,4].

Once they are nucleated, deformation twins usually propagate quickly by moving the twin boundaries across the matrix grain. This process corresponds to a stage on the stress–strain curve with near-zero strain hardening. Introducing microscopic barriers to hinder the motion of twin boundaries (TBs) is an effective way to strengthen Mg alloys [5–7]. Some studies suggested that the segregation of certain solute atoms to the TBs can exert a strong pinning effect on the migration of TBs, resulting in a significant enhancement in hardness and mechanical strength [8–10]. The solute/TB interaction can even serve as a new atomic-scale mechanism for dynamic strain aging [11]. Interactions between TBs and precipitates of various shapes (e.g., basal plates, prismatic plates, c-rods) have also been examined by both experimental and computational approaches to elucidate the strengthening effect of those precipitates [12–15]. On the other hand, little attention has been paid to the interaction between TBs and solute clusters, which are somewhere between single solute atoms and crystallized precipitates.

Clusters are recognized to form in the early stages of precipitation and have a vital influence on mechanical properties of Mg-RE alloys, such as Mg-Y and Mg-Gd alloys [16,17]. Gd-rich clusters are found to segregate onto high-angle grain boundaries, which leads to the grain refinement and texture weakening in Mg-Gd alloys [17]. Clusters of non-RE elements have been less well studied. Recently, Ca-containing Mg alloys have received strong attention because of their excellent mechanical properties and low fabrication cost [18–22]. In particular, Mg-Zn-Ca alloys demonstrate ultra-high ductility and moderate texture [23–27]. Those

features are often attributed to the co-segregation of Zn and Ca atoms onto grain boundaries (GBs), which is believed to enhance GB cohesion. On the other hand, how Zn-Ca clusters would affect the deformation mechanisms in the grain interior is not well understood. In the present work, we study the interaction between Zn-Ca clusters and TBs by molecular dynamics (MD) simulations. The result will help us better understand the work hardening behavior of Mg-Zn-Ca alloys.

2. Experimental Methods

2.1. Material Synthesis

A moderate-textured Mg-1.8Zn-0.2Ca (wt.%) alloy (denoted as ZX20) was casted and extruded. The ZX20 alloy was made from pure Mg, pure Zn, and Mg-20 wt.% Ca master alloy in an electric resistance furnace under protective gas consisting of CO_2 (99 vol.%) and SF_6 (1 vol.%). The melt was poured into a cylindrical steel mold preheated to 200 °C and then naturally cooled in air. The actual chemical composition of the cast billets was determined by an inductively coupled plasma atomic emission spectroscopy (ICP-AES) analyzer (Perkin-Elmer, Plasma 400, Norwalk, CT, USA). The cast billets were machined into cylindrical samples 60 mm in diameter and 70 mm in height, followed by homogenization at 400 °C for 12 h. One-step direct extrusion was carried out at 250 °C and 2 mm/s to produce round bars of 14 mm in diameter, which corresponds to an extrusion ratio of ~18:1.

2.2. Microstructural Characterization and Mechanical Test

Microstructures of the alloy were characterized with electron backscattered diffraction (EBSD) and transmission electron microscopy (TEM). The surface of the sample was mechanically ground using sandpapers and then electro-polished in an ethanol–10% perchloric acid electrolyte for EBSD characterization. TEM was employed to characterize finer microstructures in the alloy using a JEM-ARM200F instrument (JEOL Ltd., Tokyo, Japan). More details of EBSD and TEM analysis can be found in Refs [28,29].

Dog-bone tensile specimens with gauge dimensions of 18.0 mm × (L) × 4.8 mm (W) × 1.4 (T) were machined from the extruded alloys for standard tension tests. Cylindrical specimens with a diameter of 5 mm and a height of 7.5 mm were used for compression tests. The loading axes in tension tests were parallel to the extrusion direction while the compression tests were conducted with loading axes 0, 45, and 90 degrees from the extrusion direction. The tension and compression tests were conducted with crosshead speeds of 0.5 and 0.2 mm/min, respectively, which corresponds to a nominal strain rate of ~5×10^{-4} s^{-1} in both tests.

2.3. Experimental Observations

As the solute Ca is dilute in this alloy, precipitates such as Mg_2Ca are not readily formed. A rich profusion of solute clusters in various sizes were found to distribute uniformly in the matrix, as revealed by high-angle annular dark field-scanning transmission electron microscopy (HAADF-STEM) as shown in Figure 1. Since the brightness of individual atomic columns in the HAADF-STEM image is proportional to the square of the averaged atomic number, each bright dot in Figure 1 represents a Zn/Ca-rich column. The presence of solute clusters rather than precipitates in the cast and as-extruded ZX20 alloy is likely the result of low extrusion temperature (250 °C) and high extrusion speed (2 mm/s) which tend to suppress the dynamic precipitation.

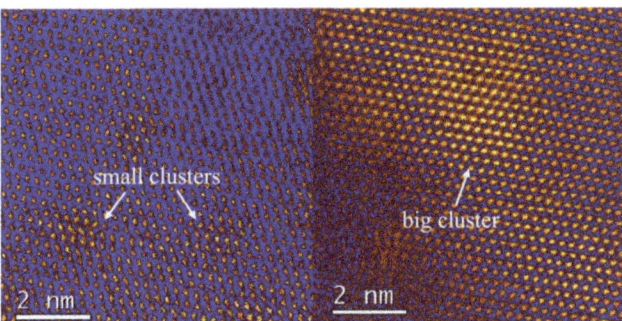

Figure 1. HAADF-STEM images showing atomic clusters in the Mg-1.8Zn-0.2Ca alloy.

For the as-extruded ZX20 alloy, a tension–compression (T-C) asymmetric behavior [30–32] is observed. As shown in Figure 2, the stress–strain curve obtained from tensile tests along the extrusion direction (ED) differs strongly from that obtained from compressive tests along the ED. The tensile yield strength (TYS ≈ 150 MPa) was higher than the compressive yield strength (CTS ≈ 100 MPa). Yet, the stress level during the compression test increased rapidly after 2% strain, and the ultimate compressive strength (UCS ≈ 500 MPa) was much higher than the ultimate tensile strength (UTS ≈ 300 MPa).

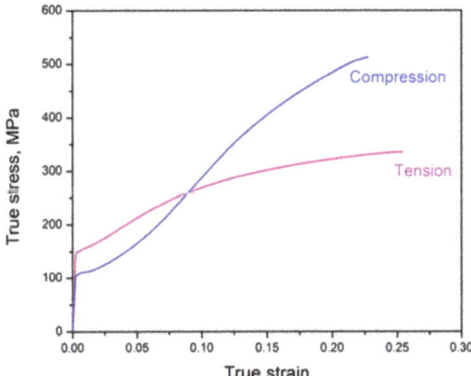

Figure 2. Stress–strain curves of the as-extruded ZX20 alloy deformed in tension and compression along ED.

The T-C asymmetry in the ZX20 alloy is attributed to the more frequent nucleation of deformation twins $\{10\bar{1}2\}$ in the compression test than that in the tension test. The ZX20 alloy exhibits a moderate extrusion texture (Figure 3). The $\{10\bar{1}2\}$ twinning would be suppressed under tension along the ED, but be favored under compression along the ED. However, compared to other as-extruded Mg alloys, such as AZ31, MN11, and Mg-5wt.%Y [33–35], the ZX20 alloy exhibits a much stronger hardening in the compressive stress–strain curve, which suggests that the growth of twins must have met some resistance. The above experimental observations motivated us to study the interaction between TBs and Zn-Ca solute clusters.

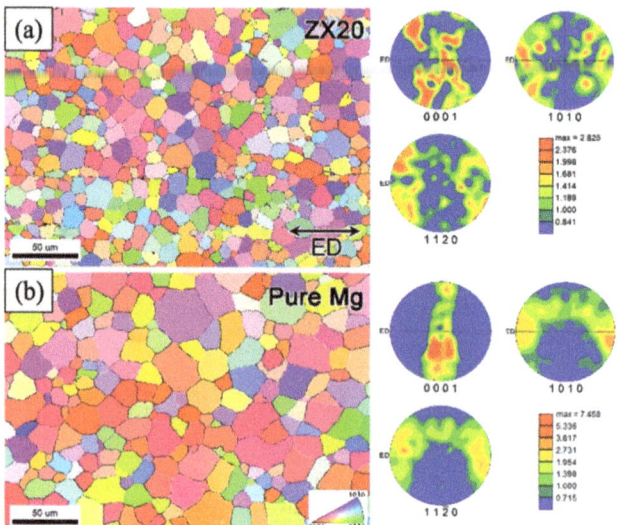

Figure 3. Inverse pole figure (IPF) maps and pole figures (PFs) of (**a**) ZX20 and (**b**) pure Mg. The colors in the IPF maps represent the grain orientations with respect to the extrusion direction (adapted from [28]).

3. Computational Procedures

MD simulations were carried out to investigate the cluster–twin interaction in the ZX20 alloy. The simulations were performed using the open-source code "LAMMPS" [36]. The modified embedded-atom method (MEAM) interatomic potential developed by Jang et al. [37] for the Mg–Zn–Ca ternary system was utilized. The visualization tool Ovito [38] was used for analyzing the simulation data, and atoms were colored by atom type as indicated in the legend of each figure containing an MD snapshot. In particular, with the common neighbor analysis (can) in Ovito, atoms at defect sites such as grain boundaries, dislocations, and faults were distinguished.

The MD sample illustrated in Figure 4 contained 821,530 atoms, having dimensions of 55 nm × 52 nm × 4.5 nm. A pair of $\{10\bar{1}2\}$ TBs at a separation distance of 32 nm were introduced within the MD supercell under three-dimensional (3D) periodic boundary conditions. The lower TB was fixed by periodically replacing Mg atoms at compression GB sites with Zn atoms such that the interaction with the cluster only involved the upper TB (the TB moving toward the cluster) [8]. To avoid additional size effects in 3D, the shapes of clusters were chosen to be cylindrical with length equal to that of the MD box in the Z direction. The cluster size was measured by the radius R of the circular cross-section. Clusters containing different Zn and Ca contents were produced by randomly replacing Mg atoms by Zn and Ca atoms at a certain Zn/Ca ratio in the cluster domain.

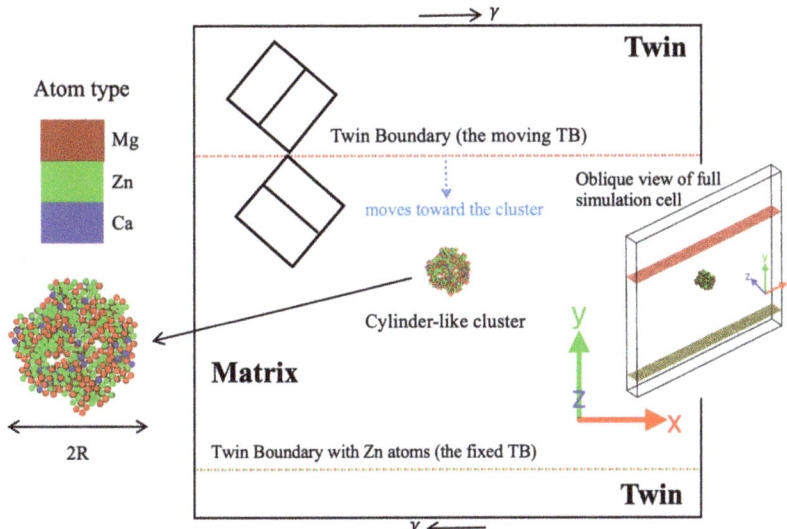

Figure 4. MD simulation supercell with a pair of horizontal TBs in Mg lattice and a Zn-Ca-containing cylinder-shaped column (cluster) residing 15 nm below the upper TB. Zn atoms (colored green) had been introduced to the lower TB sites to replace Mg atoms periodically, such that only the upper TB tended to migrate downward when the MD box was deformed by applying an X–Y plane simple shear at a constant rate. The MD supercell was $55 \times 52 \times 4.5$ nm^3 under the 3D periodic boundary conditions. Both the Zn/Ca ratio and the size of the cluster could be varied in our study. Atoms sitting at perfect hcp lattice sites are not shown.

Local strains and stresses were generated due to mismatch in atomic size which was most significant when Ca was introduced. The contents of Ca and Zn in the cluster zone were therefore limited to avoid unexpected defect nucleation, such as dislocations and stacking faults. Thus, only 30% Mg atoms of clusters were set to be replaceable in this study, and the Zn/Ca ratios and radius of clusters were chosen to be x:y $\in$ [0:10, 1:9, 5:5, 9:1, 10:0] and R $\in$ [1.5, 2.5, 4.0] nm, respectively. In this case, the Zn/Ca ratio (indicated as xZnyCa in the following parts, i.e., 10Zn, 1Zn9Ca, 9Zn1Ca, etc.) refers only to the 30% replaceable Mg atoms, irrespective of other Mg atoms.

Energy minimizations were performed on the entire supercell to relax the TB and the cluster. The isothermal–isobaric (NPT) ensemble was then employed to increase the temperature to 300 K at a time interval of 200 ps in zero-pressure conditions and the state was further relaxed at the same temperature by 100 ps with the canonical (NVT) ensemble. Then, simple shear strain was applied in the microcanonical (NVE) ensemble at a strain rate of 1×10^8/s, which was slow enough to avoid artificial kinetic effects according to previous MD simulations [39–41]. The simple shear was applied in the X–Y plane to provide a driving force in accordance with the twinning shear to move the TB in the negative Y direction towards the cluster. A timestep of 0.001 ps was used in the MD simulations and the stress, total energy, and atomic position data were stored every 1 ps. To eliminate thermal noises, all figures were exported after performing conjugate gradient energy minimizations for 5ps to eliminate thermal fluctuation of the atomic structure and atoms at perfect Mg lattice sites were hidden for better observation [42].

4. Results and Discussion
4.1. Effect of the Zn/Ca Ratio

To clarify how cluster-hardening effects depend on the Zn/Ca ratio, a set of clusters with fixed radius R = 2.5 nm were considered. The corresponding stress–strain curves obtained by MD simulations are shown in Figure 5. The cluster-free case is taken as a reference for comparison. Due to the X–Y deformation in current simulations, the shear stress is computed as the sum value of the per-atom τ_{xy} in the MD supercell, while the shear strain is defined as $\Delta L / L0$, where ΔL is the displacement distance in the X direction from the unstrained orientation and $L0$ is the box length in the Y direction. It is worth noting that since twinning is the only mode of plastic deformation due to the absence of lattice dislocations or other defects, the yield stress defines the critical resolved shear stress (CRSS) for TB migration [43,44].

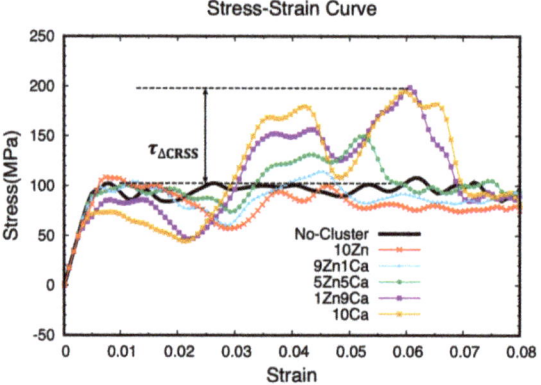

Figure 5. Shear stress–strain curves for clusters having different Zn/Ca ratios when they are sheared by the upper TB.

According to the stress–strain curves, the CRSS of TB migration is found to be ~100 MPa for the cluster-free case, and the subsequent flow stress stays at nearly the same level, indicating that there is hardly any strain hardening effect in the Mg lattice. When clusters are present, the yield points are found to differ significantly. For Zn-rich clusters (i.e., 9Zn1Ca, or 10Zn), the yield stress and the subsequent flow stress are also close to 100 MPa, suggesting a weak cluster pinning effect on TB. On the contrary, the yield stress and the subsequent flow stress apparently decreased with Ca-rich clusters (i.e., 1Zn9Ca or 10Ca), indicating that an attraction force has been imposed on the TB, which is probably caused by lattice distortions due to the large atom size of Ca. A similar attraction force was also observed between precipitates and an approaching TB in a recent study [45]. Then, immediately, a hump-like stress–strain response is observed when the TB migration is temporarily blocked by the Ca-rich cluster after their interaction. This essentially applies a hardening effect, which can be quantified by $\tau_{\Delta CRSS}$ defined as the difference between the peak flow stress in the presence of clusters and the flow stress in the cluster-free case (Figure 5). It is found that the $\tau_{\Delta CRSS}$ value increases from ~10 MPa to nearly 100 MPa as the Ca:Zn ratio rises from 1:9 to 10:0.

According to the stress–strain curves in Figure 5, a work hardening coefficient about 2000 MPa ($\theta = d\sigma/d\varepsilon$) was calculated [46–48] for the 1Zn9Ca and 10Ca cases, and about 1000 MPa for the 5Zn5Ca case. The ZX20 alloy actually has a θ value of about 750 MPa, which is comparable to the above computational values.

The interactions between a migrating TB and two clusters in different Zn/Ca ratios, 9Zn1Ca and 1Zn9Ca, are depicted in Figure 6a,b. After the onset of plastic yielding, the TB starts migrating towards the cluster by the glide of twinning dislocations, as shown for the T1 step in Figure 6, in agreement with previous MD simulations [45].

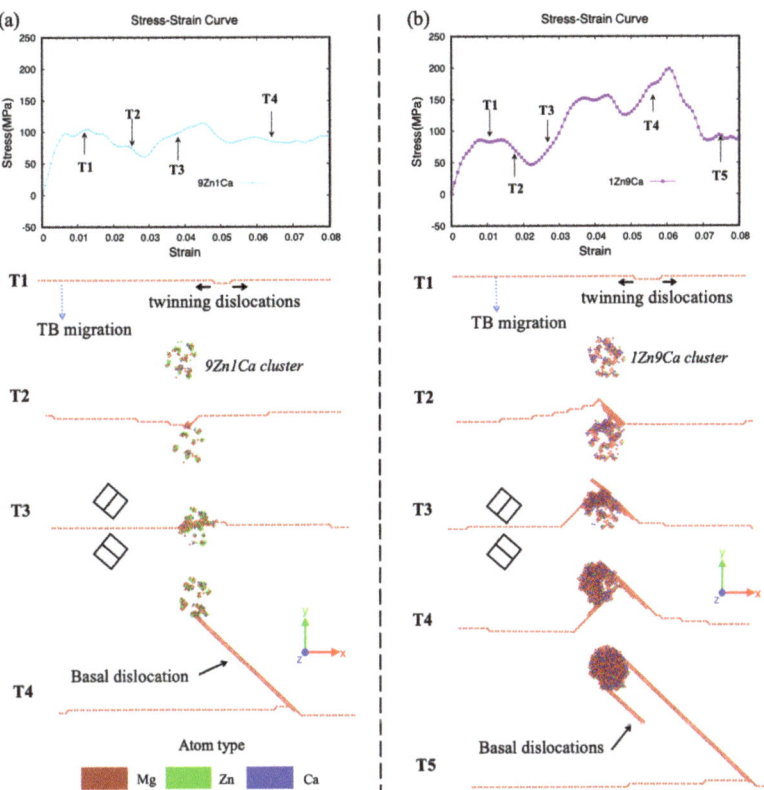

Figure 6. Snapshots showing two interaction events between a TB and a cluster having R = 2.5 nm with (**a**) Zn:Ca = 9:1 and (**b**) Zn:Ca = 1:9. Atoms sitting at perfect hcp lattice sites are not shown. These figures are sliced at Z = 0, while full interaction processes are shown in online Supplemental Movies I and II.

Afterwards, two different ways of interaction are observed. The TB can easily cut through the 9Zn1Ca cluster by leaving a basal dislocation behind, while the TB is significantly tangled with the 1Zn9Ca cluster. The different hardening effect by 9Zn1Ca and 1Zn9Ca clusters is likely a result of the misfit strain due to different atomic radii of Mg (150 pm), Zn (135 pm), and Ca (180 pm) [49]. The larger lattice distortion by the 1Zn9Ca cluster due to the higher concentration of Ca forces plenty of Mg atoms out of their original positions. In fact, more non-perfect hcp atoms are found for the 1Zn9Ca cluster, as shown in Figure 6. The misfit strain caused by the lattice distortion makes it difficult for TB to pass through the 1Zn9Ca cluster, leading to more stress.

It is interesting to notice that, after fully departing from the clusters, trailing dislocations connecting the TB and the cluster are observable. These <a> dislocations on basal planes are nucleated from the cluster surface to accommodate the high strain incompatibility between the cluster and twin, leading to a local plastic relaxation once the local stress and misfit strain are sufficiently high [50]. The number of the trailing dislocations increases when Ca dominates the cluster.

From Figure 6, the Ca-rich (1Zn9Ca) cluster has extra resistance against the migration of TBs in Mg-Zn-Ca alloys, indicating a stronger hardening effect than in the Zn-rich (9Zn1Ca) cluster.

4.2. Effect of the Cluster Size and Cluster Concentration

Size is another factor that influences the interactions of clusters with the TB. Stress–strain curves derived from TB interacting with 1Zn9Ca clusters with R = 1.5, 2.5, and 4.0 nm are shown in Figure 7a. It is found that the CRSS of TB migration dramatically reduces to 50 MPa and the $\tau_{\Delta CRSS}$ rises up to nearly 150 MPa when the cluster has R = 4.0 nm. Apparently, the attraction force becomes stronger when increasing the cluster size, and so does the hardening effect.

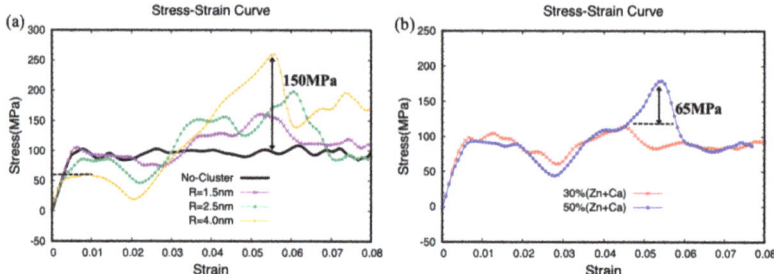

Figure 7. The stress–strain curves of cluster–TB interactions for (**a**) 1Zn9Ca clusters with a series of cluster radii R. (**b**) 9Zn1Ca clusters having R = 2.5nm with 30% and 50% (Zn + Ca), respectively.

Moreover, comparing the stress–strain curves for the 9Zn1Ca cluster with R = 2.5 nm (Figure 6a) and the 1Zn9Ca cluster with R = 1.5 nm (Figure 7a) suggests that the latter exhibits a more pronounced hardening effect. Therefore, the element content is regarded to play a more significant role than the cluster size in terms of hardening.

The effect of cluster concentration was studied as well. As mentioned earlier, the cluster considered here consisted of 70% Mg atoms and 30% (Zn + Ca) atoms to avoid unexpected defects resulting from excessive misfit strain before imposing shear strain. For the 9Zn1Ca case, however, the ratio of (Zn + Ca) can reach 50% without causing additional defects, which is due to the smaller atomic radius of Zn. The stress–strain curves in Figure 7b show that increasing the (Zn + Ca) ratio of the cluster from 30% to 50% can significantly enhance the hardening effect.

5. Conclusions

In this work, to understand the pronounced hardening effects observed experimentally, atomistic simulations are performed to systematically investigate the interactions between clusters and moving {10-12} twinning boundaries in Mg-Zn-Ca alloy. Our simulation results indicate that the cluster produced a hindering effect on TB migration, inducing a significant hardening effect in Mg-Zn-Ca alloy. The following conclusions can be reached:

(1) Increasing the Ca content can aggravate the lattice distortion of clusters and results in a stronger hardening effect. Furthermore, the cluster hardening effect is also in a positive relationship with the size and concentration of the clusters.
(2) The attractive effect of clusters on the TB is identified. The attractive force increases with the Ca content and size of the clusters.
(3) Although the twin boundary is eventually able to bypass the clusters, basal dislocations are left behind, which causes a local plastic relaxation.
(4) The Zn:Ca ratio in the cluster is found to play a more significant role than the cluster size in impeding TB migration.

Author Contributions: Conceptualization, L.W. and R.L.; methodology, R.L. and J.W.; software, R.L.; validation, R.L., Z.J. and L.W.; formal analysis, R.L.; investigation, R.L.; resources, R.L. and J.W.; data curation, R.L.; writing—original draft preparation, R.L.; writing—review and editing, Z.J. and L.W.; visualization, R.L.; supervision, L.W. and X.Z.; project administration, X.Z.; funding acquisition, X.Z. All authors have read and agreed to the published version of the manuscript.

Funding: This work is funded by the National Natural Science Foundation of China (No. 51825101) and the Shanghai Rising-Star Program (No. 20QA1405000). MD calculations were carried out using the cluster resource (π2.0) provided by the HPC Center, Shanghai Jiao Tong University.

Data Availability Statement: Not applicated.

Conflicts of Interest: The authors declare no conflict of interest.

References

1. Bettles, C.; Barnett, M. Introduction. In *Advances in Wrought Magnesium Alloys*; Bettles, C., Barnett, M., Eds.; Woodhead Publishing: Melbourne, Australia, 2012; pp. 12–13, ISBN 978-1-84569-968-0.
2. Pollock, T.M. Weight Loss with Magnesium Alloys. *Science* **2010**, *328*, 986. [CrossRef] [PubMed]
3. Christian, J.W.; Mahajan, S. Deformation Twinning. *Prog. Mater. Sci.* **1995**, *39*, 1–157. [CrossRef]
4. Wang, J.; Hoagland, R.G.; Hirth, J.P.; Capolungo, L.; Beyerlein, I.J.; Tomé, C.N. Nucleation of a (-1012) Twin in Hexagonal Close-Packed Crystals. *Scr. Mater.* **2009**, *61*, 903–906. [CrossRef]
5. Suh, B.-C.; Shim, M.-S.; Shin, K.S.; Kim, N.J. Current Issues in Magnesium Sheet Alloys: Where Do We Go from Here? *Scr. Mater.* **2014**, *84–85*, 1–6. [CrossRef]
6. Kim, N.J. Critical Assessment 6: Magnesium Sheet Alloys: Viable Alternatives to Steels? *Mater. Sci. Technol.* **2014**, *30*, 1925–1928. [CrossRef]
7. Suh, B.-C.; Kim, J.H.; Bae, J.H.; Hwang, J.H.; Shim, M.-S.; Kim, N.J. Effect of Sn Addition on the Microstructure and Deformation Behavior of Mg-3Al Alloy. *Acta Mater.* **2017**, *124*, 268–279. [CrossRef]
8. Nie, J.F.; Zhu, Y.M.; Liu, J.Z.; Fang, X.Y. Periodic Segregation of Solute Atoms in Fully Coherent Twin Boundaries. *Science* **2013**, *340*, 957. [CrossRef]
9. Xin, Y.; Zhang, Y.; Yu, H.; Chen, H.; Liu, Q. The Different Effects of Solute Segregation at Twin Boundaries on Mechanical Behaviors of Twinning and Detwinning. *Mater. Sci. Eng. A* **2015**, *644*, 365–373. [CrossRef]
10. Pei, Z.; Li, R.; Nie, J.F.; Morris, J.R. First-Principles Study of the Solute Segregation in Twin Boundaries in Mg and Possible Descriptors for Mechanical Properties. *Mater. Des.* **2019**, *165*, 107574. [CrossRef]
11. Hooshmand, M.S.; Ghazisaeidi, M. Solute/Twin Boundary Interaction as a New Atomic-Scale Mechanism for Dynamic Strain Aging. *Acta Mater.* **2020**, *188*, 711–719. [CrossRef]
12. Vaid, A.; Guénolé, J.; Prakash, A.; Korte-Kerzel, S.; Bitzek, E. Atomistic Simulations of Basal Dislocations in Mg Interacting with Mg17Al12 Precipitates. *Materialia* **2019**, *7*, 100355. [CrossRef]
13. Liao, M.; Li, B.; Horstemeyer, M.F. Interaction between Prismatic Slip and a Mg17Al12 Precipitate in Magnesium. *Comput. Mater. Sci.* **2013**, *79*, 534–539. [CrossRef]
14. Fan, H.; Zhu, Y.; El-Awady, J.A.; Raabe, D. Precipitation Hardening Effects on Extension Twinning in Magnesium Alloys. *Int. J. Plast.* **2018**, *106*, 186–202. [CrossRef]
15. Tang, X.Z.; Guo, Y.F. The Engulfment of Precipitate by Extension Twinning in Mg–Al Alloy. *Scr. Mater.* **2020**, *188*, 195–199. [CrossRef]
16. Nie, J.F.; Wilson, N.C.; Zhu, Y.M.; Xu, Z. Solute Clusters and GP Zones in Binary Mg-RE Alloys. *Acta Mater.* **2016**, *106*, 260–271. [CrossRef]
17. Bugnet, M.; Kula, A.; Niewczas, M.; Botton, G.A. Segregation and Clustering of Solutes at Grain Boundaries in Mg-Rare Earth Solid Solutions. *Acta Mater.* **2014**, *79*, 66–73. [CrossRef]
18. Guo, F.; Pei, R.; Jiang, L.; Zhang, D.; Korte-Kerzel, S.; Al-Samman, T. The Role of Recrystallization and Grain Growth in Optimizing the Sheet Texture of Magnesium Alloys with Calcium Addition during Annealing. *J. Magnes. Alloy.* **2020**, *8*, 252–268. [CrossRef]
19. Zeng, Z.R.; Bian, M.Z.; Xu, S.W.; Davies, C.H.J.; Birbilis, N.; Nie, J.F. Effects of Dilute Additions of Zn and Ca on Ductility of Magnesium Alloy Sheet. *Mater. Sci. Eng. A* **2016**, *674*, 459–471. [CrossRef]
20. Pan, H.; Yang, C.; Yang, Y.; Dai, Y.; Zhou, D.; Chai, L.; Huang, Q.; Yang, Q.; Liu, S.; Ren, Y.; et al. Ultra-Fine Grain Size and Exceptionally High Strength in Dilute Mg–Ca Alloys Achieved by Conventional One-Step Extrusion. *Mater. Lett.* **2019**, *237*, 65–68. [CrossRef]
21. Zhang, A.; Kang, R.; Wu, L.; Pan, H.; Xie, H.; Huang, Q.; Liu, Y.; Ai, Z.; Ma, L.; Ren, Y.; et al. A New Rare-Earth-Free Mg-Sn-Ca-Mn Wrought Alloy with Ultra-High Strength and Good Ductility. *Mater. Sci. Eng. A* **2019**, *754*, 269–274. [CrossRef]
22. Pan, H.; Kang, R.; Li, J.; Xie, H.; Zeng, Z.; Huang, Q.; Yang, C.; Ren, Y.; Qin, G. Mechanistic Investigation of a Low-Alloy Mg–Ca-Based Extrusion Alloy with High Strength–Ductility Synergy. *Acta Mater.* **2020**, *186*, 278–290. [CrossRef]
23. Wasiur-Rahman, S.; Medraj, M. Critical Assessment and Thermodynamic Modeling of the Binary Mg-Zn, Ca-Zn and Ternary Mg-Ca-Zn Systems. *Intermetallics* **2009**, *17*, 847–864. [CrossRef]

24. Zeng, Z.R.; Zhu, Y.M.; Bian, M.Z.; Xu, S.W.; Davies, C.H.J.; Birbilis, N.; Nie, J.F. Annealing Strengthening in a Dilute Mg-Zn-Ca Sheet Alloy. *Scr. Mater.* **2015**, *107*, 127–130. [CrossRef]
25. Liu, C.; Chen, X.; Chen, J.; Atrens, A.; Pan, F. The Effects of Ca and Mn on the Microstructure, Texture and Mechanical Properties of Mg-4 Zn Alloy. *J. Magnes. Alloy.* **2020**, *9*, 1084–1097. [CrossRef]
26. Nandy, S.; Tsai, S.P.; Stephenson, L.; Raabe, D.; Zaefferer, S. The Role of Ca, Al and Zn on Room Temperature Ductility and Grain Boundary Cohesion of Magnesium. *J. Magnes. Alloy.* **2021**, *9*, 1521–1536. [CrossRef]
27. Mostaed, E.; Sikora-Jasinska, M.; Wang, L.; Mostaed, A.; Reaney, I.M.; Drelich, J.W. Tailoring the Mechanical and Degradation Performance of Mg-2.0Zn-0.5Ca-0.4Mn Alloy Through Microstructure Design. *JOM* **2020**, *72*, 1880–1891. [CrossRef]
28. Wang, J.; Zhu, G.; Wang, L.; Vasilev, E.; Park, J.-S.; Sha, G.; Zeng, X.; Knezevic, M. Origins of High Ductility Exhibited by an Extruded Magnesium Alloy Mg-1.8Zn-0.2Ca: Experiments and Crystal Plasticity Modeling. *J. Mater. Sci. Technol.* **2021**, *84*, 27–42. [CrossRef]
29. Wang, J.; Zhu, G.; Wang, L.; Zhu, Q.; Vasilev, E.; Zeng, X.; Knezevic, M. Dislocation-Induced Plastic Instability in a Rare Earth Containing Magnesium Alloy. *Materialia* **2021**, *15*, 101038. [CrossRef]
30. Song, B.; Pan, H.; Ren, W.; Guo, N.; Wu, Z.; Xin, R. Tension-Compression Asymmetry of a Rolled Mg-Y-Nd Alloy. *Met. Mater. Int.* **2017**, *23*, 683–690. [CrossRef]
31. Nie, J.F.; Shin, K.S.; Zeng, Z.R. Microstructure, Deformation, and Property of Wrought Magnesium Alloys. *Metall. Mater. Trans. A* **2020**, *51*, 6045–6109. [CrossRef]
32. Tong, L.B.; Zheng, M.Y.; Kamado, S.; Zhang, D.P.; Meng, J.; Cheng, L.R.; Zhang, H.J. Reducing the Tension-Compression Yield Asymmetry of Extruded Mg-Zn-Ca Alloy via Equal Channel Angular Pressing. *J. Magnes. Alloy* **2015**, *3*, 302–308. [CrossRef]
33. Hidalgo-Manrique, P.; Robson, J.D.; Pérez-Prado, M.T. Precipitation Strengthening and Reversed Yield Stress Asymmetry in Mg Alloys Containing Rare-Earth Elements: A Quantitative Study. *Acta Mater.* **2017**, *124*, 456–467. [CrossRef]
34. Dobroň, P.; Hegedüs, M.; Olejňák, J.; Drozdenko, D.; Horváth, K.; Bohlen, J. Influence of Thermomechanical Treatment on Tension–Compression Yield Asymmetry of Extruded Mg–Zn–Ca Alloy. In *Magnesium Technology*; Joshi, V., Jordon, J., Orlov, D., Neelameggham, N., Eds.; The Minerals, Metals & Materials Series; Springer: Cham, Switzerland, 2019. [CrossRef]
35. Yin, D.D.; Boehlert, C.J.; Long, L.J.; Huang, G.H.; Zhou, H.; Zheng, J.; Wang, Q.D. Tension-Compression Asymmetry and the Underlying Slip/Twinning Activity in Extruded Mg-Y Sheets. *Int. J. Plast.* **2021**, *136*, 102878. [CrossRef]
36. Plimpton, S. Fast Parallel Algorithms for Short-Range Molecular Dynamics. *J. Comput. Phys.* **1995**, *117*, 1–19. [CrossRef]
37. Jang, H.-S.; Seol, D.; Lee, B.-J. Modified Embedded-Atom Method Interatomic Potentials for Mg–Al–Ca and Mg–Al–Zn Ternary Systems. *J. Magnes. Alloy* **2021**, *9*, 317–335. [CrossRef]
38. Stukowski, A. Visualization and Analysis of Atomistic Simulation Data with OVITO-the Open Visualization Tool. *Model. Simul. Mater. Sci. Eng.* **2010**, *18*, 015012. [CrossRef]
39. Luque, A.; Ghazisaeidi, M.; Curtin, W.A. Deformation Modes in Magnesium (0 0 0 1) and (011-1) Single Crystals: Simulations versus Experiments. *Model. Simul. Mater. Sci. Eng.* **2013**, *21*, 045010. [CrossRef]
40. Dou, Y.; Luo, H.; Zhang, J. The Effects of Yttrium on the {10-12} Twinning Behaviour in Magnesium Alloys: A Molecular Dynamics Study. *Philos. Mag. Lett.* **2020**, *100*, 224–234. [CrossRef]
41. Esteban-Manzanares, G.; Ma, A.; Papadimitriou, I.; Martínez, E.; Llorca, J. Basal Dislocation/Precipitate Interactions in Mg-Al Alloys: An Atomistic Investigation. *Model. Simul. Mater. Sci. Eng.* **2019**, *27*, 075003. [CrossRef]
42. Fan, H.; Wang, Q.; Tian, X.; El-Awady, J.A. Temperature Effects on the Mobility of Pyramidal <c+a> dislocations in Magnesium. *Scr. Mater.* **2017**, *127*, 68–71. [CrossRef]
43. El Kadiri, H.; Barrett, C.D.; Wang, J.; Tomé, C.N. Why Are {101¯2} Twins Profuse in Magnesium? *Acta Mater.* **2015**, *85*, 354–361. [CrossRef]
44. Wang, J.; Beyerlein, I.J.; Tomé, C.N. Reactions of Lattice Dislocations with Grain Boundaries in Mg: Implications on the Micro Scale from Atomic-Scale Calculations. *Int. J. Plast.* **2014**, *56*, 156–172. [CrossRef]
45. Fan, H.; Zhu, Y.; Wang, Q. Effect of Precipitate Orientation on the Twinning Deformation in Magnesium Alloys. *Comput. Mater. Sci.* **2018**, *155*, 378–382. [CrossRef]
46. Trojanová, Z.; Drozd, D.; Halmešová, K.; Džugan, J.; Škraban, T.; Minárik, P.; Németh, G.; Lukáč, P. Strain Hardening in an AZ31 Alloy Submitted to Rotary Swaging. *Materials* **2021**, *14*, 157. [CrossRef]
47. Balík, J.; Dobroň, P.; Chmelík, F.; Kužel, R.; Drozdenko, D.; Bohlen, J.; Letzig, D.; Lukáč, P. Modeling of the Work Hardening in Magnesium Alloy Sheets. *Int. J. Plast.* **2016**, *76*, 166–185. [CrossRef]
48. Máthis, K.; Trojanová, Z.; Lukáč, P. Hardening and Softening in Deformed Magnesium Alloys. *Mater. Sci. Eng. A* **2002**, *324*, 141–144. [CrossRef]
49. Slater, J.C. Atomic Radii in Crystals. *J. Chem. Phys.* **1964**, *41*, 3199–3204. [CrossRef]
50. Robson, J.D.; Stanford, N.; Barnett, M.R. Effect of Precipitate Shape on Slip and Twinning in Magnesium Alloys. *Acta Mater.* **2011**, *59*, 1945–1956. [CrossRef]

MDPI
St. Alban-Anlage 66
4052 Basel
Switzerland
Tel. +41 61 683 77 34
Fax +41 61 302 89 18
www.mdpi.com

Metals Editorial Office
E-mail: metals@mdpi.com
www.mdpi.com/journal/metals